Maynooth Library
00364715

Signals and Systems

D0714688

TUTORIAL GUIDES IN ELECTRONIC ENGINEERING

Series editors
Professor G.G. Bloodworth, *University of York*
Professor A.P. Dorey, *University of Lancaster*
Professor J.K. Fidler, *University of York*

This series is aimed at first- and second-year undergraduate courses. Each text is complete in itself, although linked with others in the series. Where possible, the trend towards a 'systems' approach is acknowledged, but classical fundamental areas of study have not been excluded. Worked examples feature prominently and indicate, where appropriate, a number of approaches to the same problem.

A format providing marginal notes has been adopted to allow the authors to include ideas and material to support the main text. These notes include references to standard mainstream texts and commentary on the applicability of solution methods, aimed particularly at covering points normally found difficult. Graded problems are provided at the end of each chapter, with answers at the end of the book.

1. Transistor Circuit Techniques: discrete and integrated (2nd edition) — G.J. Ritchie
2. Feedback Circuits and Op Amps (2nd edition) — D.H. Horrocks
3. Pascal for Electronic Engineers (2nd edition) — J. Attikiouzel
4. Computers and Microprocessors: components and systems (2nd edition) — A.C. Downton
5. Telecommunication Principles (2nd edition) — J.J. O'Reilly
6. Digital Logic Techniques: principles and practice (2nd edition) — T.J. Stonham
7. Instrumentation: Transducers and Interfacing (new edition) — B.R. Bannister and D.G. Whitehead
8. Signals and Systems: models and behaviour (2nd edition) — M.L. Meade and C.R. Dillon
9. Basic Electromagnetism and its Applications — A.J. Compton
10. Electromagnetism for Electronic Engineers — R.G. Carter
11. Power Electronics — D.A. Bradley
12. Semiconductor Devices: how they work — J.J. Sparkes
13. Electronic Components and Technology: engineering applications — S.J. Sangwine
14. Optoelectronics — J. Watson
15. Control Engineering — C. Bissell
16. Basic Mathematics for Electronic Engineers: models and applications — Szymanski
17. Software Engineering — D. Ince
18. Integrated Circuit Design and Technology — M.J. Morant

621.38 MEA

Signals and Systems

Models and behaviour

Second edition

M.L. Meade and C.R. Dillon
Faculty of Technology
The Open University

KLUWER ACADEMIC PUBLISHERS
DORDRECHT / BOSTON / LONDON

1065218
07-3251
ELEC

Library of Congress Cataloging-in-Publication Data
Meade, M.L.
Signals and systems: models and behaviour/M.L. Meade and C.R. Dillon. — 2nd ed.
p. cm. — (Tutorial guides in electronic engineering; 8)
Includes bibliographical references and index.
ISBN 0-412-40110-X. — ISBN 0-442-31460-4 (USA)
1. Signal processing—Mathematics. 2. System analysis. 3. Transformations (Mathematics) I. Dillon, C.R. (Christopher Richard) II. Title. III. Series.
TK5102.5.M365 1991
621.382'2—dc20 91-12210
CIP

ISBN 0-412-40110-X 0-442-31460-4 (USA)

Published by Kluwer Academic Publishers,
P.O. Box 17, 3300 AA Dordrecht, The Netherlands.

Sold and distributed in the North, Central and South America
by Kluwer Academic Publishers,
101 Philip Drive, Norwell, MA 02061, U.S.A.

In all other countries, sold and distributed
by Kluwer Academic Publishers Group,
P.O. Box 322, 3300 AH Dordrecht, The Netherlands.

Printed on acid-free paper

First edition 1986
Reprinted 1987, 1990
Second edition 1991
Reprinted 2000

All Rights Reserved
© 1986, 1991, 2000 M.L. Meade and C.R. Dillon
No part of the material protected by this copyright notice may be reproduced or utilized in any form or by any means, electronic or mechanical, including photocopying, recording or by any information storage and retrieval system, without written permission from the copyright owner

Printed in the Netherlands.

Contents

Preface

This book was written for first and second year undergraduates in electronic engineering and the physical sciences with the aim of providing a firm grounding in the study of signals and systems. We have made few assumptions about entry behaviour beyond a familiarity with first year circuit principles and mathematics and our treatment is broadly based to enable progression to more specialized courses in telecommunications, control engineering and signal processing.

To this end we have chosen to deal with both transform techniques and pole-zero descriptions and we have made a special effort to bring out the essential unity of continuous-time and discrete-time methods of analysis and representation. This reflects the growing and welcome trend to integrate teaching in these important subject areas which have for so long been treated as separate and distinct topics.

Our general approach has been greatly influenced by our earlier collaboration on the Open University undergraduate course T326 Electronic Signal Processing and we are grateful to colleagues in the Electronics Discipline at the Open University for many hours of fruitful discussions. We also wish to thank series editors Professor Tony Dorey and Professor Kel Fidler for their comments and guidance.

Preface to second edition

In preparing the second edition of *Signals and Systems*, we have given further thought to the order of presentation of key concepts and taken the opportunity to provide a more detailed treatment of the z-transform. A new, final, chapter serves to review and integrate much of the earlier material, while introducing the Discrete Fourier Transform in the context of signal capture and spectral analysis.

These changes reflect the growing importance of discrete-time topics in undergraduate studies and the fact that many students now have direct experience in the use and application of signal-processing software. As far as possible, the emphasis is placed on discussion rather than on detailed mathematical proofs, with the aim of developing and encouraging a critical approach to the use of software tools.

Introduction to Signals and Systems 1

Objectives

- ☐ To explain what is meant by the terms signal and system and to distinguish between continuous-time and discrete-time operation.
- ☐ To show how discrete-time signals may be formed by sampling a continuous-time or analogue signal.
- ☐ To introduce the idea of a system model.
- ☐ To define the properties of time-invariant linear systems.
- ☐ To explain what is meant by stability and causality.
- ☐ To distinguish between some important categories of signal.
- ☐ To introduce some basic signal-processing operations giving the average value and power of a signal and to explain the terms 'energy signal' and 'power signal'.
- ☐ To define the properties of signals with symmetry.
- ☐ To explain what is meant by orthogonal signals.
- ☐ To introduce the idea of interpolating signal samples and to show how aliasing errors result from inadequate sampling.
- ☐ To introduce the sampling theorem.

A signal can be regarded as the variation of any measurable quantity that conveys information concerning the behaviour of a related system. Whether we are dealing with engineering systems, biological systems, transport, communication or economic systems, we rely on the interpretation of signal records, charts, graphs and displays to increase understanding, to make decisions and, when appropriate, to verify that a particular system is performing to specification.

Effective interpretation of a signal requires careful judgement and a knowledge of the various *signal processing* techniques that might be applied to reveal its information content. It may be decided, for example, to rid a signal of short-term fluctuations which obscure its underlying long-term behaviour. A more complex series of operations could be defined to separate a wanted signal from an intrusive interfering signal. Further processing might then be applied to determine the peak value, average value, average power, rate of change or any other aspect of a signal that may be considered significant and yet might not be apparent on first inspection.

Signal classification

Signals come in many shapes and forms and can be represented in any number of ways; as continuous chart recordings for example, as bar charts, tables of values, patterns of magnetic domains on a computer disc or as patterns of dots on a computer display. It will be noted that the manner of representation chosen will be different according to whether the signal is a function of a single variable, such as time or length, or a function of several variables, such as time and horizontal and vertical position. In this book, we shall be dealing only with signals that depend on a single variable and it will be assumed – without any real loss of generality – that the independent variable is *time*.

As we shall now explain, time-dependent signals can be divided into two main classes. The classification depends on the characteristics of the time variable which may be *continuous* or *discrete*.

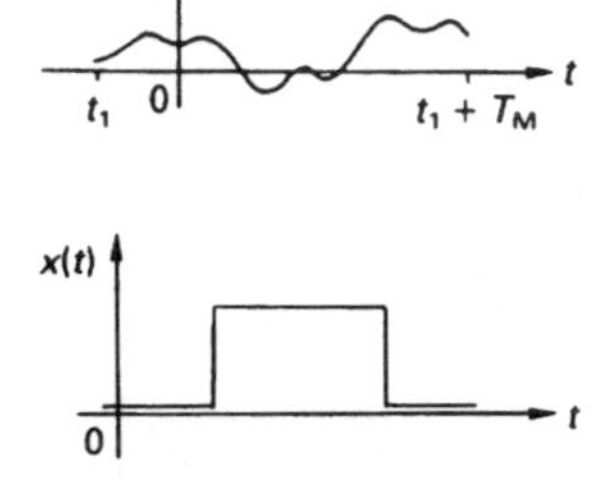

The pulse signal is discontinuous in amplitude but continuous in time.

Continuous-time signals

A *continuous-time* signal can be represented mathematically as a function of a continuous time variable. The graph of a continuous-time signal $x(t)$ is thus defined at each and every instant over a measurement interval extending from $t = t_1$ to $t = t_1 + T_M$. The signal opposite is an example of a continuous-time signal that takes on values in a continuous range. Most naturally occurring signals, such as speech and music signals, fall into this category and are known as continuous-time *analogue* signals. It is important to emphasize that a continuous-time signal is not necessarily described by a mathematically continuous function. An example of a continuous-time signal with a discontinuous waveform is the pulse signal sketched in the margin.

Temp

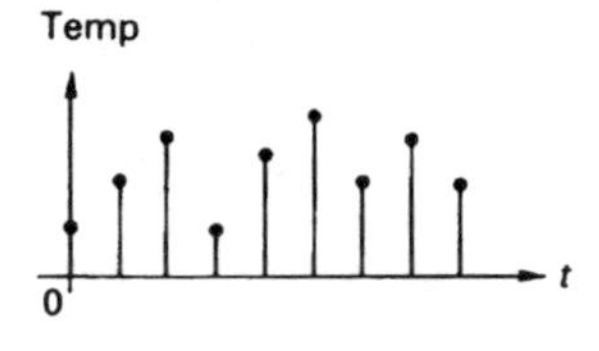

We have no knowledge of a discrete-time signal other than at the sampling instants.

Discrete-time signals

A *discrete-time* signal is defined only at a particular set of instants in time. The values of a discrete-time signal could represent midday temperatures, monthly rainfall or annual profits. In any of these cases the signal graph will have the appearance indicated in the marginal figure.

The instants at which the data values are plotted are called the *sampling instants*. It will be assumed throughout that the sampling instants are equally spaced so that the discrete-time signal is defined for $t = nT$, $n = 0, \pm 1, \pm 2, \ldots$, where n is an integer and T is known as the *sampling interval*.

Since the sampling interval remains constant in any particular application, we can adopt the index n of the discrete-time instants as the independent variable. This leads to the representation $x[n]$ and $y[n]$, in which the values of a discrete-time signal are expressed as functions of the integer variable, n. For the purposes of analysis and processing, $x[n]$ can be regarded as an *ordered sequence* of numbers with values $x[0], x[1], x[2], \ldots$ for $n = 0, 1, 2, \ldots$

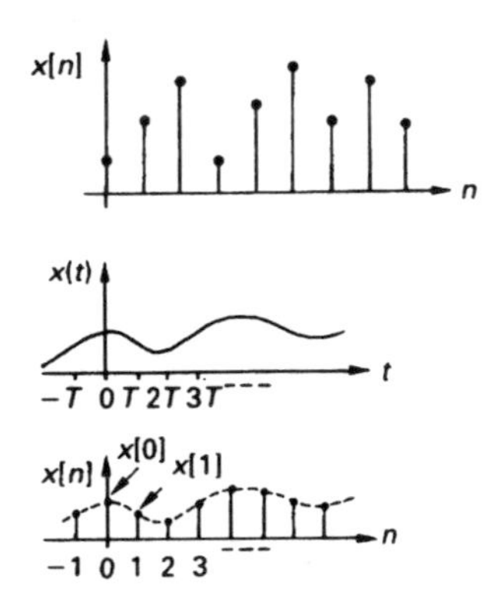

In many practical applications, the values of a discrete-time signal or sequence $x[n]$ are obtained by *sampling* a continuous-time signal $x(t)$ at equidistant points along the time axis. If the sampling interval is T, the sequence $x[n]$ is defined by

$$x[n] = x(nT), \; n = 0, \pm 1, \pm 2, \ldots$$

This process is known as *periodic sampling*. The number of samples taken per second is equal to $f_s = 1/T$, where f_s is called the *sampling frequency*. The idea that we can choose the sampling frequency in such a way that $x[n]$ provides a complete specification of $x(t)$ will be discussed later, at the end of this chapter.

Worked Example 1.1

A sinusoidal signal $x(t) = \sin 2\pi t$ is sampled at a frequency $f_s = 8$ Hz beginning at time $t = 0$. Write down the first few values of the sample sequence.

Solution: The sampling interval $T = 1/f_s = 1/8$ s. The sample sequence is:

$$x[n] = \sin 2\pi nT = \sin n\pi/4, \text{ for } n = 0, 1, 2, 3, \ldots$$

We obtain the values $x[0] = 0$, $x[1] = \sqrt{2}/2$, $x[2] = 1$, $x[3] = \sqrt{2}/2, \ldots$, giving the sequence $x[n] = 0, \sqrt{2}/2, 1, \sqrt{2}/2, 0, -\sqrt{2}/2, \ldots$ When we plot the graph of a discrete-time sequence, it is essential that we show the zero values. By convention, all undefined values are assumed to be zero.

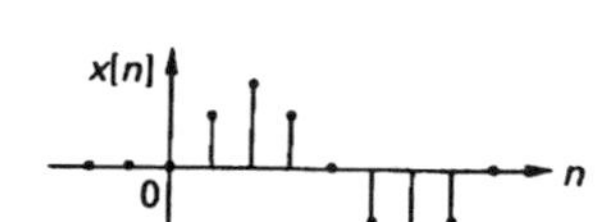

Signal processing systems

We have used the term *system* in its most general sense and given some examples of highly complex systems in which cause and effect relationships are barely understood. In the chapters which follow, however, we shall be concerned only with systems with well-defined properties which may be used to process signals in some desired way. The most simple representation of a *signal processor* is a 'black box' with a single input and a single output. We suppose that a signal, x, applied to the input gives rise to a *response*, y, at the output and represent this input–output relationship by writing

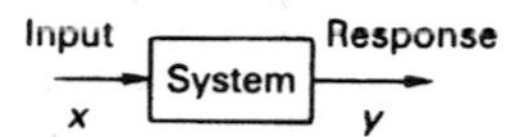

In a continuous-time system the input and output signals are continuous functions of time, $x(t)$ and $y(t)$. In a discrete-time system the input and output are sequences $x[n]$ and $y[n]$.

$$\begin{array}{ccc} \text{Input} & \rightarrow & \text{Response} \\ x & & y \end{array}$$

where the arrow denotes a *causal connection*.

The term *causal connection* implies that the input x causes or gives rise to the output y.

We can describe the nature of the signal processing operation by treating the black box as a mathematical rather than as a physical entity in which the input and output signals are connected by mathematical equations rather than by physical components. These equations are known generally as the *model* of the system. The model might be arrived at by *analysing* an existing system or it may form the basis for system design or *synthesis* where we seek to specify a set of components and their interconnections to carry out a particular signal processing function.

Continuous-time signal processors

A *continuous-time signal processor* is a system in which both the input and the output are continuous-time signals. The equations relating the input and output of a continuous-time processor are the equations of ordinary algebra and calculus and involve operations such as continuous integration and differentiation of signals, addition and subtraction of signals and signal squaring.

Worked Example 1.2

A continuous-time signal $x(t)$ represents the variation of *displacement* in a mechanical system. Specify the behaviour of a continuous-time system that converts $x(t)$ to a *velocity* signal.

Solution: Velocity is defined to be the rate of change of displacement. The required conversion could therefore be provided by an analogue system modelled by the input–output equation

$$y(t) = \frac{dx(t)}{dt}$$

The system model in the previous example defines the *function* of the system but tells us nothing about its physical content. The task of a system designer is to produce a working system that approximates the behaviour of the mathematical model. In practice this could take the form of an analogue electronic circuit comprising a connection of resistors, capacitors and operational amplifiers.

Discrete-time signal processors

Hardware and software aspects of digital systems and the topic of binary coding are covered in Stonham, T.J., *Digital Logic Techniques*, Downton, A.C., *Computers and Microprocessors* and Attikiouzel, J., *Pascal for Electronic Engineers*, all in the Tutorial Guides in Electronic Engineering series, Chapman and Hall.

For many years all analysis and processing of signals in electrical form was carried out using analogue electronic equipment. During the last two decades, however, there has been a steady shift towards the use of digital computers and dedicated digital hardware devices in signal processing applications. At first, cost, size and speed considerations weighed in favour of the analogue-based approach. Nowadays the situation is quite the reverse; digital VLSI devices are available for specialist processing tasks while there is a wealth of signal processing software developed for use with personal computers.

In digital signal processing the input and output signals are represented by *sequences* of numerical values. It will be appreciated that in a digital processor the sequence values are encoded in (usually) binary form. A *digital* sequence thus differs from an ordinary *discrete-time* sequence $x[n]$ or $y[n]$ in that its values are *quantized*: they take on a finite number of possible values that depends on the number of *bits* used in the encoding. However, when analysing systems and devising signal processing routines, it is often found convenient to ignore this distinction and to regard all sequences as discrete-time signals. When this is done the processing system can be modelled and represented as a *discrete-time signal processor*, that is a system which operates on an input discrete-time signal or sequence $x[n]$ and produces, in response, a sequence $y[n]$ at the output.

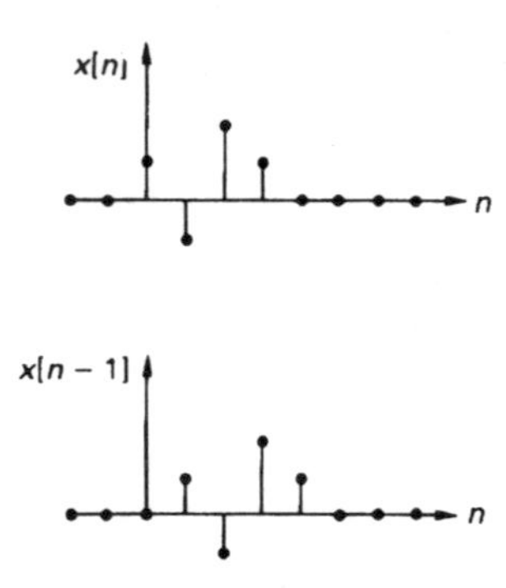

Whereas it is very difficult to delay an analogue signal by an appreciable amount without introducing distortion in the signal waveform, we can operate repeatedly on a sequence to introduce delays of any order, measured in multiples of the sampling interval, T.

We shall now indicate the range and type of operations that can be carried out using a discrete-time system. We refer first to the diagram opposite which shows two sequences labelled $x[n]$ and $x[n-1]$. The notation $x[n-1]$ signifies that $x[n]$ has been *delayed* by one sampling interval. The graph of the delayed sequence is obtained by shifting the graph of the original sequence $x[n]$ by one unit to the *right* along the horizontal axis. A discrete-time system that performs a delay operation is represented symbolically by the *delay* block, labelled T in Figure 1.1a. If this operation is repeated m times we generate the sequence $x[n-m]$, which represents $x[n]$ delayed by m sampling intervals.

Figure 1.1 shows a number of other operations that can be performed by a

(a) Change of scale: $y(t) = kx(t)$ (k is a constant)
(b) Integration: $y(t) = \int x(t)\mathrm{d}t$
(c) Squaring: $y(t) = x^2(t)$

Solution:
(a) *Change of scale*: An input $x(t)$ gives a response $kx(t)$. If we put $x(t) = x_1(t) + x_2(t)$, where $x_1(t)$ and $x_2(t)$ are two input components, the response will be

$$kx(t) = k[x_1(t) + x_2(t)] = kx_1(t) + kx_2(t).$$

In other words, the total response is given by the sum of the individual responses to the component signals, and the operation is *linear*.
(b) *Integration*: The causal connection between input and response is

$$x(t) \to \int x(t)\,\mathrm{d}t.$$

Putting $x(t) = x_1(t) + x_2(t)$, we obtain the response

$$\int x(t)\,\mathrm{d}t = \int [x_1(t) + x_2(t)]\,\mathrm{d}t = \int x_1(t)\,\mathrm{d}t + \int x_2(t)\,\mathrm{d}t.$$

Again, the total response is the sum of the responses to the individual input components. Hence the operation of integration is *linear*.
(c) *Squaring*: We have input $x(t) \to$ response $x^2(t)$
Putting $x(t) = x_1(t) + x_2(t)$ gives

$$x^2(t) = [x_1(t) + x_2(t)]^2 = x_1^2(t) + 2x_1(t)x_2(t) + x_2^2(t).$$

In this case the overall response is *not* equal to the sum of the individual responses; hence squaring a signal is a *non-linear* operation.

Exercise 1.2 The following processing operations are performed by a discrete-time system:

(a) averaging, $y[n] = \frac{1}{N}\sum_{n=0}^{N-1} x[n]$

(b) square-rooting, $y[n] = \sqrt{x[n]}$.

Classify each operation as linear or non-linear.

It is important to remember that linearity, homogeneity and time-invariance are properties of our *models*. We must maintain a clear distinction between the models we use in pencil-and-paper design studies and the physical systems that we actually build and use. However it is usually possible to achieve an acceptable match only over a limited range of operation. In a signal amplifier, for example, the maximum output voltage that can be produced is limited by the power supply voltage. In this case the principle of homogeneity cannot hold if increasing the input signal by a factor a leads to a value for the output which exceeds the maximum output voltage. Similarly the principle of superposition cannot hold if the addition of two or more signal components results in amplifier output limiting. In digital systems the results of arithmetic processing operations are stored in finite-length registers. To ensure that the

A great deal of time and engineering skill is devoted to making the behaviour of physical systems match the idealized properties of our models as closely as possible.

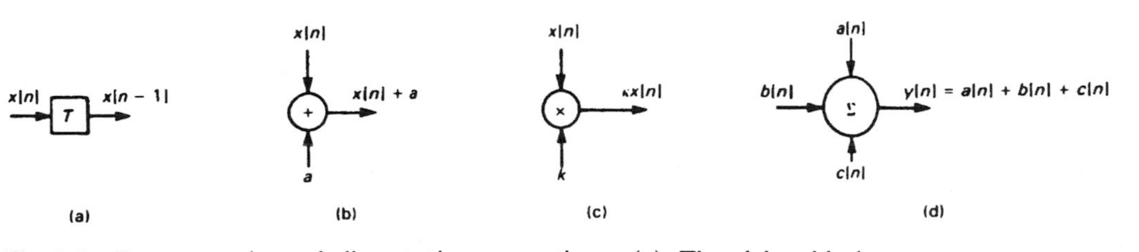

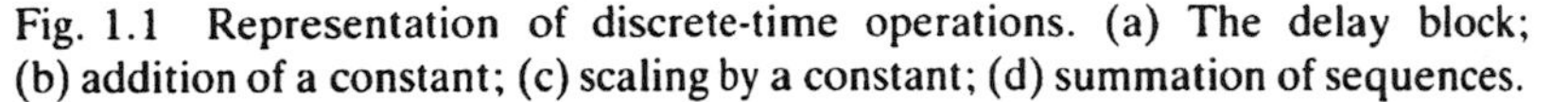

Fig. 1.1 Representation of discrete-time operations. (a) The delay block; (b) addition of a constant; (c) scaling by a constant; (d) summation of sequences.

discrete-time signal processor. In each case the operation is shown within a circle and we use arrows to indicate the direction of flow of the signals. The *addition* block in Figure 1.1b is used to add a constant a to each of the values of a sequence $x[n]$, while we use the *scaling* or *multiplier* block in Figure 1.1c to indicate that each and every value in a sequence is multiplied by a constant k. The *summation* block in Figure 1.1d can have as its inputs two or more *sequences*, $a[n]$, $b[n]$, $c[n]$ and so on. Its effect is to add the values of $a[n]$ to the corresponding values of the other input sequences. The sequence of output values is thus

$$y[0] = a[0] + b[0] + c[0],$$
$$y[1] = a[1] + b[1] + c[1],$$

with the nth value of the output sequence given by

$$y[n] = a[n] + b[n] + c[n].$$

We can use the same notation $x[n]$ to represent an entire discrete-time signal or sequence or simply to denote the nth value of the sequence. This is similar to the idea that the function $x(t)$ represents a continuous-time signal and also gives the value of x at time t.

Exercise 1.1 A discrete-time sequence is defined by $x[n] = n^2$ for $n \geq 0$. Write down the first few values of the output sequence when $x[n]$ is applied to a) an addition block with $a = 1$, and b) a multiplier block with $k = 2$.

Discrete-time processors can carry out many of the input–output operations that were traditionally the preserve of analogue systems. The following example demonstrates how we can specify the operation of a discrete-time system in a particular application.

Worked Example 1.3 Show how the operations carried out by the continuous-time system in the previous example could be approximated by taking samples of the displacement signal $x(t)$.

Solution: For a sufficiently small interval T, the derivative of a signal $x(t)$ can be approximated by

$$y(t) \simeq \frac{x(t) - x(t - T)}{T}.$$

A discrete-time system can simulate the operation of an analogue system. But it should be emphasized that discrete-time systems can also be used to process signals in ways that cannot be achieved using traditional analogue techniques. We should regard discrete-time systems as being worthy of study in their own right rather than as approximations to analogue systems.

If we take samples of the input signal $x(t)$ at times $t = nT$, we obtain the sequence of values, $x[n] = x(nT)$. To find the samples of $x(t - T)$ it is not necessary to repeat the sampling process. All we have to do is to take the existing sequence $x[n]$ and shift the sample values by one sampling interval to obtain the sequence $x[n - 1]$. Thus:

$$x[n] = x(nT),$$
$$x[n - 1] = x(nT - T).$$

We can now operate on these two sequences of samples to generate a third sequence $y[n]$, which approximates the samples of the derivative signal:

$$y[n] \simeq \frac{x[n] - x[n - 1]}{T}.$$

$$y[n] = \frac{x[n] - x[n - 1]}{T}$$

We can represent the operations specified in the previous example by drawing a system diagram. Since input and output are simply strings of numbers, these arithmetic operations could be carried out by using a suitably programmed digital computer or by using purpose-built hardware. We shall not be concerned with specific hardware or software implementations however. Our interest is solely in the system model which shows the discrete-time system as a set of interconnected summation, multiplication and delay elements.

Linearity and time-invariance

In this book we shall be concerned almost exclusively with a particular class of system known as *time-invariant linear systems*. The properties and descriptions of such systems have been studied extensively by engineers, physicists and mathematicians for a long time. One reason for this interest is that we find that the behaviour of many different types of physical system – electrical, mechanical, thermal and so on – may be modelled successfully using linear theory. Many analogue systems can be modelled quite adequately in terms of linear differential equations. Although differential equations are not central to our approach here, we shall spend some time with their discrete-time counterpart, the *linear difference equation* which will be introduced in Chapter 2. The first step meanwhile must be to define the rules governing the behaviour of linear models.

A time-invariant linear system can be described by a linear differential equation having constant (time-invariant) coefficients.

Linearity

We define a *linear* system as one which obeys the principle of *superposition*:

> If an input x_1 causes an output y_1 and an input x_2 causes an output y_2 then a system is linear if and only if an input $x_1 + x_2$ causes an output $y_1 + y_2$

It is often convenient to model the input signal as a sum of *components*. In this case we can work out the response of a linear system to each component separately and then use superposition to find the total response as the sum of the component responses.

Homogeneity

If an input x_1 to a linear system produces a response y_1 then, using superposition, an input $x_1 + x_1 = 2x_1$ will produce a response $2y_2$. In more general terms we can say that if x_1 produces y_1,

$$x_1 \rightarrow y_1$$

then a scaled version of the input ax_1 will produce a similarly scaled output ay_1,

In our system models the principle holds irrespective of whether the constant *a* is a real, imaginary or complex number.

$$ax_1 \rightarrow ay_1$$

where a is a constant. This result is formally called the principle of *homogeneity*.

Time-invariance

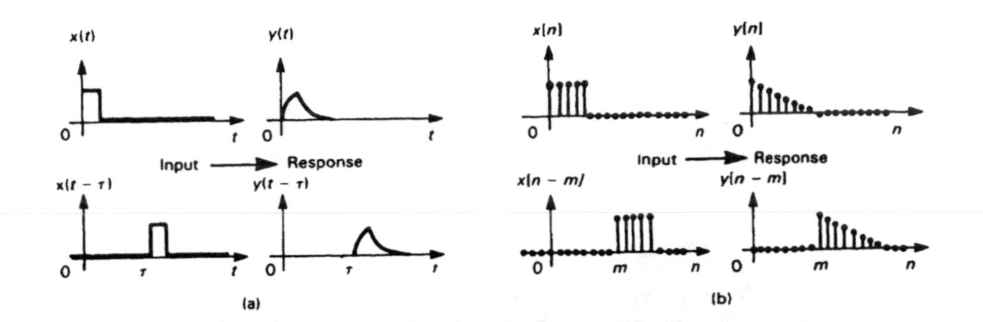

Fig. 1.2 Response of a time-invariant system to a delayed input signal. (a) Continuous-time; (b) discrete-time.

Except for some advanced applications which are beyond the scope of this book we usually assume that the properties of our systems do not change with time. To see what this means in practice, suppose that an input signal $x(t)$ causes an output $y(t)$ as in Figure 1.2a. Next, we consider the system response to a signal $x(t - \tau)$ which is a *delayed* version of $x(t)$. If the system is *time-invariant* then regardless of the value of τ, a delayed input $x(t - \tau)$ will produce a delayed response $y(t - \tau)$. Similarly, for discrete-time signals, if

The notation $x(t - \tau)$ implies that $x(t)$ has been shifted to the *right* on the time axis by τ seconds.

$$x[n] \rightarrow y[n]$$

then if the input sequence is delayed for m sample intervals the resulting output sequence will be delayed by the same amount:

Remember that the notation $x[n - m]$ implies that a discrete-time signal $x[n]$ has been delayed by m sample intervals.

$$x[n - m] \rightarrow y[n - m]$$

Figure 1.2b illustrates the effect. Notice that in both cases the delayed response is identical in shape and size to the original response. Thus, built into our models is the assumption that the response of a system to a given input will always be the same irrespective of when the input is applied.

Worked Example 1.4

Classify the following continuous-time signal-processing operations as linear or non-linear:

principles of superposition and homogeneity hold the data must be suitably scaled if operations such as addition and multiplication are not to result in numbers that lie outside the range that can be handled by the processor.

Stability and causality

Two important ideas underlie our discussion of linearity; these are the concepts of *stability* and *causality*. Suppose, for example, that we apply a sinewave to a linear system as part of frequency-response test and observe the response of the system. If we now remove the input we expect to see the response die away with time. Eventually, we would expect the system to settle back to a state where both the input and the output are zero. A system is said to be *stable* if it has the property of 'forgetting' about an input when that input is removed, and returning to a quiescent state of its own accord. In contrast, the response of an *unstable* system continues for an indefinite time after the input has ceased, often in the form of an exponentially increasing output. In a practical system the response will continue to grow only until some physical limitation (such as the supply voltage in an electrical system) is reached. The system is then in a new state with different rules governing its behaviour and the original properties of linearity no longer hold.

A system must be stable if we are to express the output as a function of, and causally related to, the input.

There is no causal relationship between the input and the output in an unstable system.

The assumption of causality is simply a statement of our general experience that effect cannot precede cause. In other words the response of a physical system to an input cannot occur before that input has been applied.

In view of these remarks we shall now define a *causal signal* to be one which takes zero values for $t < 0$. The samples of a causal signal will thus be a *causal sequence* $x[n]$ with the property $x[n] = 0$ for $n < 0$.

Although we can easily construct systems that continue to produce an output after the input has been removed, no actual system can be expected to 'predict' the future by responding before the input occurs.

A causal sequence $x[n]$ is applied to a discrete-time *integrator* as illustrated. The output is initially zero. Show that after $n + 1$ inputs the output will have the value

$$y[n] = x[0] + x[1] + x[2] + \ldots + x[n].$$

Consider what happens if the input is removed and comment on the stability of this arrangement.

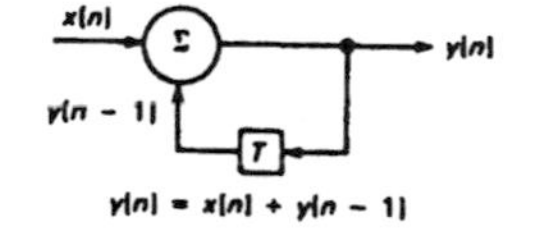

Signal types and definitions

A mathematical expression that describes or approximates the variation of a signal with time is known as a *time-domain* model. A sequence $x[n]$ could provide a time-domain model for a signal $x(t)$ but it serves to represent the signal only at the sampling instants $t = nT$. A continuous-time model on the other hand is defined at every instant within the interval occupied by the signal; for example the expression

$$x(t) = A\cos\omega t, \text{ for } t_1 < t < t_1 + T_M$$

representing the variation of a sinusoidal signal.

A simple signal like the sinusoid can be represented by a compact and convenient mathematical expression that can be held to be valid over arbitrarily long intervals. More complicated signals of course will have models of corresponding complexity and an important aim of *signal analysis* is to generate such models in a form that is both understandable and applicable. One way to do this is to represent a given signal as a sum of elementary *components* and we shall describe a number of workable approaches in the following chapters, applicable to both continuous- and discrete-time signals.

We can use the principle of superposition to find the response of a linear system to any signal expressed as a sum of components.

Given that a signal is either continuous-time or discrete-time, the first step in analysis is usually to place the signal in one of the categories described in the following sections. This is a necessary step because we shall find that the choice of analysis technique is dependent largely on the nature of the signal involved. While most of the following discussion will be given in terms of continuous-time signals it should be stressed that the conclusions apply equally to their discrete-time counterparts.

Periodic and aperiodic signals

The values of a time-dependent *periodic signal* repeat every T_0 seconds where T_0 is known as the *period* of the signal. Many signals of natural origin can be modelled as having periodic behaviour and it is also common practice to generate periodic signals in the laboratory for the purpose of system testing and evaluation.

Figure 1.3 shows examples of periodic *test* signals having sinusoidal, triangular and pulse form.

A mathematical model describing the waveform of a continuous-time periodic signal has the property

$$x(t) = x(t + T_0), \text{ for all } t. \tag{1.1}$$

Strictly speaking, the period of $x(t)$ is the smallest value of T_0 for which this relationship holds. $\cos \omega\tau$ and $\sin \omega\tau$ both have a period $T_0 = 2\pi/\omega$ but they also repeat over any interval of length kT_0, where k is an integer.

The model is thus defined for all times past, present and future, whereas an observable signal – even a carefully controlled test signal – will be subject to time-dependent distortions and irregularities. The idea behind the model therefore, is to represent the essential features of a periodic signal and not to imply that its waveform is replicated exactly and indefinitely over the entire time axis.

If we sample a periodic signal at precisely N samples per period, the resulting sample sequence $x[n]$ will be periodic with the values recurring every N units: $x[n] = x[n + N]$.

If a signal is periodic then we can in principle choose a measurement interval of sufficient length to display several cycles of its waveform. If a signal is *aperiodic* however, its waveform will not repeat within the measurement interval and fails to do so even if the interval is made arbitrarily long. Figure 1.4 shows contrasting versions of aperiodic behaviour. First of all, examples of *pulse-like* phenomena where the signal is either of limited duration or

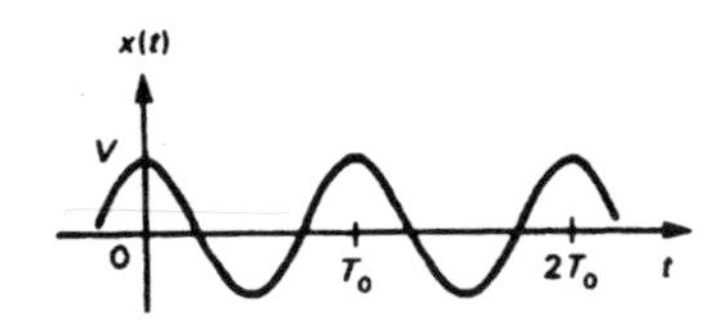

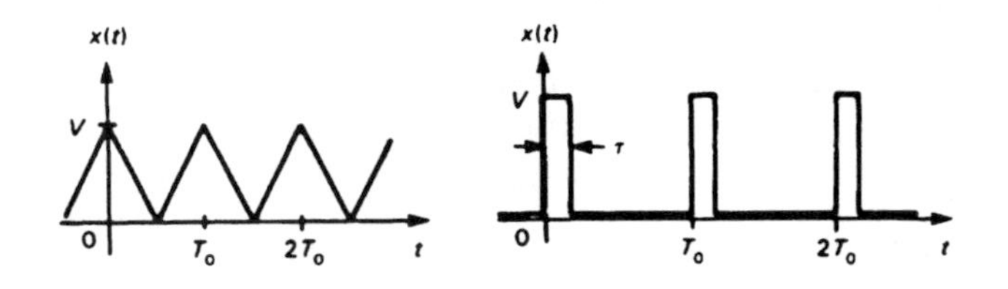

Fig. 1.3 Examples of periodic test signals.

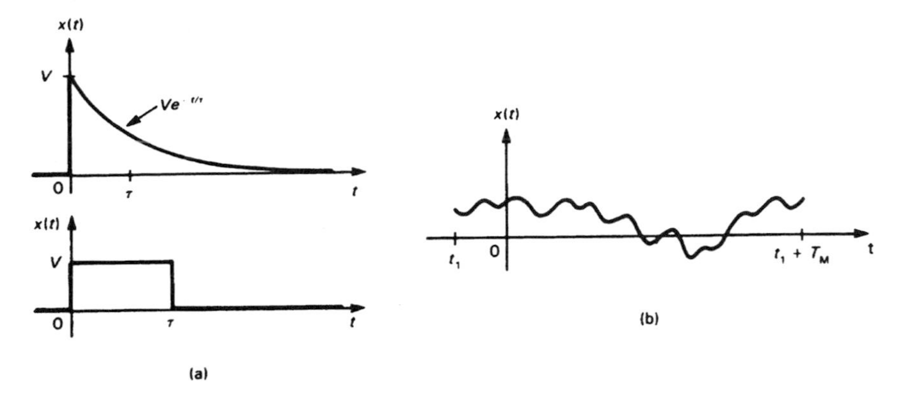

Fig. 1.4 Examples of aperiodic signals. (a) Pulse-like signals; (b) a randomly fluctuating signal.

eventually decays to zero. Secondly, in Figure 1.4b, a fluctuating signal that does not die away with time and which is available for measurement over arbitrarily long intervals. The difficulty in this case is to infer the properties of the entire signal, working from signal records of finite duration. A study of the properties of fluctuating signals and, indeed, all signals with a *random* content brings us into the area of *probability* and *statistics* which lies beyond the scope of this book. The best that we can do here will be to cover the first stages in the analysis process where we begin with a finite signal record or, alternatively, a set of samples, and then begin to analyse the signal as though it were *deterministic*, with no uncertainty about its values.

The properties of random signals and the statistical approach to signal analysis are reviewed in O'Reilly, J.J., *Telecommunication Principles*, Van Nostrand Reinhold, 1984.

The values of a deterministic signal model are defined for all time.

Average value

The graph of a time-dependent signal varies above and below a well-defined level known as the *average* or *mean* value x_0, as shown here. For a continuous-time signal $x(t)$, recorded over an interval $t = t_1$ to $t = t_1 + T_M$, the average value is given by the signal processing operation

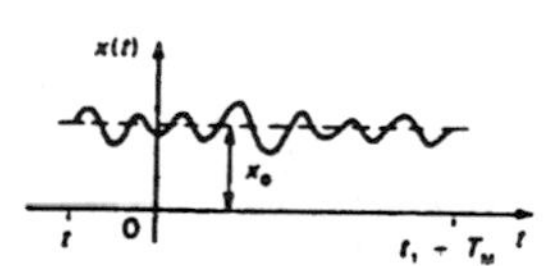

$$x_0 = \frac{1}{T_M}\int_{t_1}^{t_1+T_M} x(t)\,dt. \tag{1.2}$$

The operation defined by Equation 1.2 is essentially *linear* and could be performed by a linear processing system.

We note in particular that the average value of a sinusoidal signal is zero over any interval containing a whole number of periods:

$$\frac{1}{kT_0}\int_{t_1}^{t_1+kT_0} A\sin\omega t\,dt = 0, \text{ for } k = 0, 1, 2, \ldots$$

where the period $T_0 = 2\pi/\omega$.

If it is not feasible to evaluate Equation 1.2 for a given signal, the average value can often be estimated with surprising accuracy 'by eye', taking into account the variation of the signal over the entire record. Alternatively, a better estimate can be obtained by taking a series of N samples over the range of the signal. The signal average is then given approximately by the average of the sample values:

This operation can be carried out using a system based on the discrete-time integrator described in Exercise 1.3.

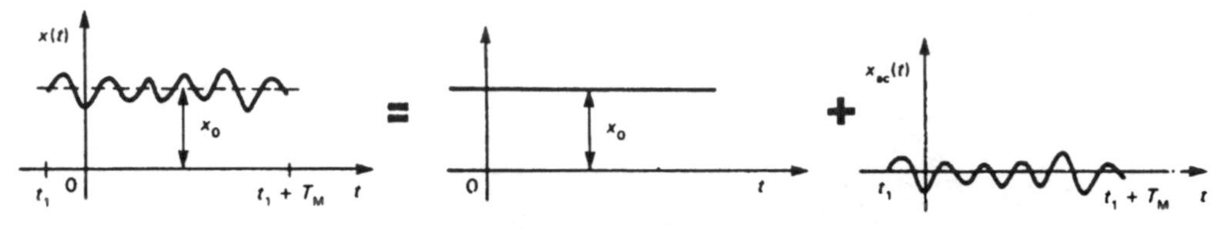

Fig. 1.5 Representation of a signal $x(t)$ as a sum of direct and alternating components.

$$x_0 \simeq \frac{1}{N}\sum_{n=0}^{N-1} x[n].$$

According to this definition, $x_{ac}(t)$ will have an average value of zero.

If we extract the average value from a signal, then we are left with a time-varying signal $x_{ac}(t)$ defined by

$$x_{ac}(t) = x(t) - x_0.$$

In this context, we call x_0 the *direct* component and $x_{ac}(t)$ the *alternating* component of the signal. In principle, any signal record of finite length can be represented as the sum of direct and alternating components as shown in Figure 1.5.

If we now consider the behaviour of a pulse-like signal such as those shown earlier in Figure 1.4, it is not difficult to show that the average value gradually falls to zero as we evaluate Equation 1.2 for longer and longer measurement intervals T_M. For example, if we take the rectangular pulse and choose a measurement interval $t = 0$ to $t = T_M$ where T_M is greater than the pulse width τ, we obtain

$$x_0 = \frac{V\tau}{T_M}.$$

The averaging operation is not normally used in the analysis of pulse-like signals.

The average value becomes very small as T_M is made much larger than the pulse duration. Thus, if we attempt to model a pulse-like signal over an arbitrarily long time interval we find that the average value and, hence, the direct component is vanishingly small.

In contrast, if a signal does not die away with time, it may be assumed to take values over any measurement interval that we care to choose. For a *periodic* signal, a 'long' measurement interval is one that includes many cycles of the signal and we can show that the *long-term* average is simply equal to the average value taken over a single period. The idea of a direct component is thus a useful one for periodic signals, and is given by the operation

$$x_0 = \frac{1}{T_0}\int_{t_1}^{t_1+T_0} x(t)\,dt$$

where T_0 is the period of the signal.

Exercise 1.4 Find the average values of the periodic pulse and triangular signals illustrated in Figure 1.3 and hence represent each signal as the sum of direct and alternating components.

Energy signals and power signals

We define an *energy signal* to be one for which the *total energy* is finite. This is the value given by the integral

$$E_{\text{tot}} = \int_{-\infty}^{\infty} |x(t)|^2 \, dt. \tag{1.3}$$

The use of the magnitude $|x(t)|$ enables this result to be extended to signal models involving complex quantities. If a signal model takes only *real* values, then $x^2(t)$ and $|x(t)|^2$ will be identical.

The energy in a signal is thus defined to be the *area* contained by the graph of the squared magnitude of the signal and the time axis and is a wholly positive quantity. Pulse-like signals that have finite duration or decay exponentially to zero have finite energy as the following example shows.

Worked Example 1.5

Find the total energy associated with the exponentially decaying pulse

$$\begin{aligned} v(t) &= V e^{-t/\tau} \quad \text{for } t \geq 0 \\ &= 0 \quad\quad\quad \text{for } t < 0. \end{aligned}$$

Solution: The signal is causal, taking zero values for $t < 0$, so the energy is given by

$$E_{\text{tot}} = V^2 \int_0^{\infty} e^{-2t/\tau} \, dt = \frac{V^2 \tau}{2}.$$

According to this definition, the energy of a voltage signal will have the dimensions [voltage]^2 · [time].

If we attempt to evaluate Equation 1.3 for a periodic signal or for any other signal that does not die away with time, the energy accumulates as longer and longer stretches of signal are taken into account and, in the limit, the total energy is infinite. To deal with these signals then, we must work with the signal *power*. For a voltage signal $v(t)$, developed across a resistance R, the *instantaneous power* is defined to be the quantity

$$P(t) = \frac{|v(t)|^2}{R} \text{ watts.}$$

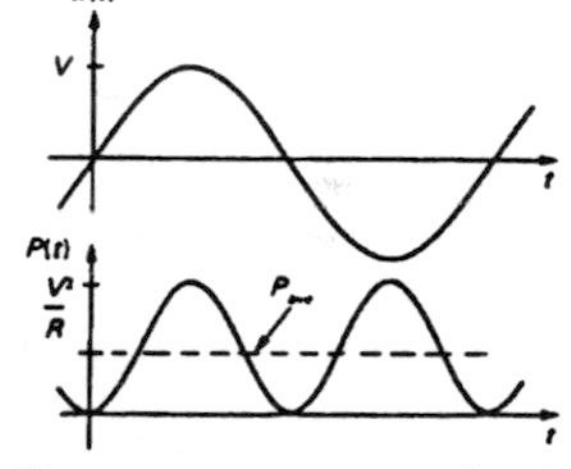

The graph of instantaneous power is everywhere positive and it can be averaged over any time interval of interest to give the *average power* delivered to the load resistance:

$$P_{\text{av.}} = \frac{1}{T_M} \int_{t_1}^{t_1 + T_M} P(t) \, dt = \frac{1}{T_M} \int_{t_1}^{t_1 + T_M} |v(t)|^2 / R \, dt.$$

The average power associated with a signal represented by a sequence of N samples may be approximated by forming the summation

$$P_{\text{av}} \simeq \frac{1}{N} \sum_{0}^{N-1} \frac{x^2[n]}{R}$$

Signals which do not die away with time are usually known as *power signals* because their average power is finite even for very long measurement intervals. In particular, for periodic signals, the average power may be found by averaging $P(t)$ over a very long interval containing many periods of the signal or by taking an average over a single period. For a periodic signal with period T_0 therefore,

$$P_{\text{av.}} = \frac{1}{T_0} \int_{t_1}^{t_1 + T_0} \frac{|v(t)|^2}{R} \, dt.$$

Worked Example 1.6 Find the average power associated with the sinusoidal signal $v(t) = V\cos\omega t$.

Solution: The instantaneous power is given by

$$P(t) = \frac{|V\cos\omega t|^2}{R} = \frac{V^2}{2R} + \frac{V^2}{2R}\cos 2\omega t.$$

Even though the dimensions may not appear to be correct, the division by an appropriate resistance is implied in this convention.

If we average $P(t)$ over any interval of length $T_0 = 2\pi/\omega$, the time-dependent component averages to zero and we obtain the average power $V^2/2R$ W. Now, the usual convention in signal analysis is to calculate power on the assumption of a 1 Ω resistive load. On this basis, the average power of the sinusoid becomes $V^2/2$ which is identical to the *mean-square value*, obtained by averaging the square of the signal.

The distinction between power signals and energy signals is an important one because it indicates the circumstances in which time averages may be used in signal analysis rather than direct integration. Thus, we shall find that the study of pulse-like energy signals involves the use of time integrals, whereas periodic signals are usually associated with the averaging operation.

Signal symmetry and orthogonality

In some textbooks r is known as a *correlation coefficient* which is used to measure the degree of similarity between two signals.

We shall find that the analysis of signals involves the repeated use of a particular operation giving the integral or average of a signal *product*, for example:

$$r = \int_{t_1}^{t_1+T_M} x_1(t)x_2(t)\,dt.$$

We can almost always save work in signal analysis by looking at the *shape* and *symmetry* of waveforms.

The result of this operation, r, will be a constant for a given pair of signals. However, in many cases r will be zero and we can often see this *before* we attempt to evaluate the integral. We shall therefore identify a number of such examples here, in line with the principle that we should avoid doing unnecessary and often tedious integrations.

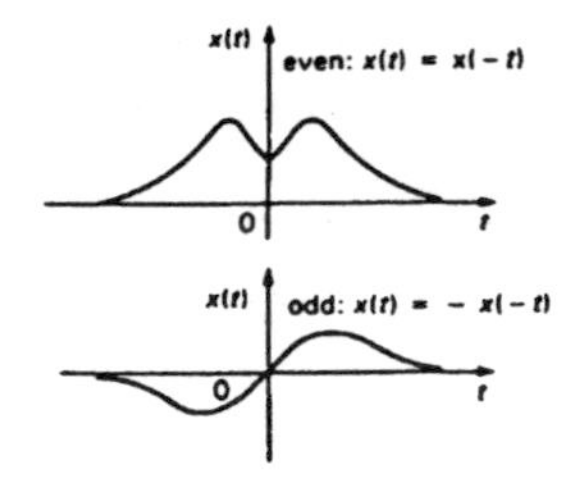

Odd and even signals

A signal $x(t)$ is said to be *even* if $x(t) = x(-t)$ and *odd* if $x(t) = -x(-t)$. The graph of an even signal is thus symmetric about $t = 0$ and the graph of an odd signal is antisymmetric. This means that an odd function will integrate to zero over any interval $t = -T$ to $t = T$, whereas for an even signal we have the relationship

$$\int_{-T}^{T} x(t)\,dt = 2\int_{0}^{T} x(t)\,dt, \text{ for } x(t) \text{ even.}$$

Worked Example 1.7 Evaluate the integral

$$\int_{-T}^{T} \cos\omega t \sin\omega t\,dt.$$

Solution: We have $\cos\omega t = \cos(-\omega t)$ (even) and $\sin\omega t = -\sin(-\omega t)$ (odd). Now the *products* of even and odd signals obey the rules:

$$\begin{aligned}(\text{even})(\text{even}) &= \text{even};\\ (\text{even})(\text{odd}) &= \text{odd};\\ (\text{odd})(\text{odd}) &= \text{even}.\end{aligned}$$

These rules correspond to the familiar rules

$$(+1)(+1) = +1;\quad (+1)(-1) = -1;\quad (-1)(-1) = +1$$

The product of $\cos\omega t$ and $\sin\omega t$ is therefore *odd* and if we integrate the product over $t = -T$ to $t = T$ the result is zero.

In fact, $\cos\omega t \sin\omega t = \frac{1}{2}\sin 2\omega t$ which is clearly an odd function.

Exercise 1.5

Figure 1.6 shows further examples of odd and even pairs. Verify that the following results hold for all values of ω by inspecting the symmetry of the waveforms rather than by evaluating the integrals

$$\int_{-T}^{T} p(t)\cos\omega t\,dt = 0;\quad \int_{-T}^{T} v(t)\sin\omega t\,dt = 0.$$

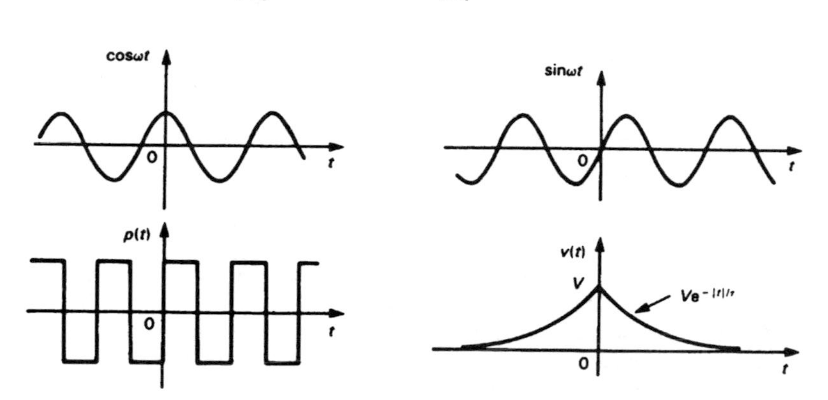

Fig. 1.6 Odd and even pairs.

Orthogonality

When the product of two signals integrates or averages to zero over a specified interval $t = t_1$ to $t = t_1 + T_M$, the signals are said to be *orthogonal* in that interval. Thus for orthogonal signals $x_1(t)$ and $x_2(t)$:

$$r = \int_{t_1}^{t_1+T_M} x_1(t)x_2(t)\,dt = 0.$$

According to this definition, $\cos\omega t$ and $\sin\omega t$ are orthogonal for $-T < t < T$. Moreover, because these signals are periodic with period $T_0 = 2\pi/\omega$ they are also orthogonal over any interval $t = t_1$ to $t = t_1 + kT_0$, where k is an integer:

The term 'orthogonal' means literally 'at right angles' and reflects the relationship between the phasors of $\cos\omega t$ and $\sin\omega t$ on a phasor diagram.

$$\int_{t_1}^{t_1+T_0} \cos\omega t \sin\omega t\,dt = 0,\ T_0 = 2\pi/\omega,\ k = 1, 2, 3, \ldots$$

As a rule, if the product of two signals is odd within a specified interval then the signals will be orthogonal within that interval. Thus the pairs of signals considered in the previous exercise can be regarded as orthogonal pairs. More generally, we can often identify orthogonal pairs simply by inspecting their waveforms and in practice it is rarely necessary to do any integration at all, especially when working with simple signal models.

Exercise 1.6

Sketch the waveform of the signal product and look for the values of ω for which the product integrates to zero.

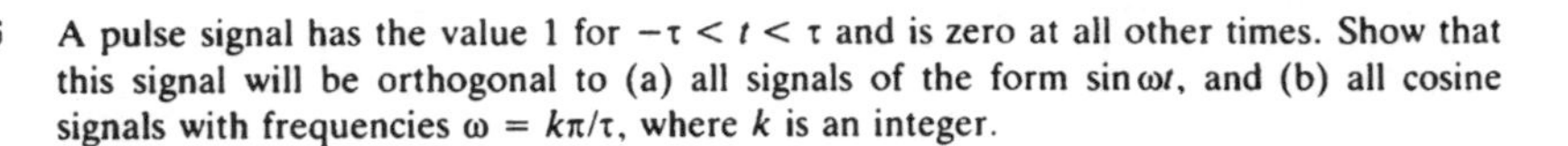

A pulse signal has the value 1 for $-\tau < t < \tau$ and is zero at all other times. Show that this signal will be orthogonal to (a) all signals of the form $\sin \omega t$, and (b) all cosine signals with frequencies $\omega = k\pi/\tau$, where k is an integer.

Signal sampling

We recall that f_s is called the sampling frequency and T the sampling interval.

The basis of signal sampling is that a continuous-time signal $x(t)$ can be represented by a sequence of samples $x[n]$ with values $x(nT)$. We shall follow the sampling scheme introduced at the beginning of this chapter, and suppose that $x[n]$ is derived from $x(t)$ by periodic sampling at a frequency $f_s = 1/T$.

A digital-to-analogue converter is a form of electronic interpolator and will usually be found at the output of a discrete-time processor to generate a continuous signal $y(t)$ from the output samples.

When we sample a signal we usually aim for a 'complete' representation which implies that we can recover the waveform of $x(t)$ by a process known as *interpolation*. This means that for practical purposes we can reproduce the waveform of the continuous-time signal by 'joining up the points' on a graph of the samples, using a smooth curve as shown in Figure 1.7a. In practice, the choice of sampling frequency for a particular signal depends on how rapidly the signal is changing with time. For example, the sample spacing will be typically less than a microsecond for a signal containing components up to 1 MHz or so, while a spacing of several seconds might be sufficient for a signal that varies on a time-scale of minutes. If the sampling frequency has been chosen correctly, then we can always recover the signal waveform by interpolation. Otherwise, if we choose too low a sampling rate as in Figure 1.7b, we lose vital information about the rapidly changing parts of the signal. When we interpolate the samples therefore, we can have no confidence in the result

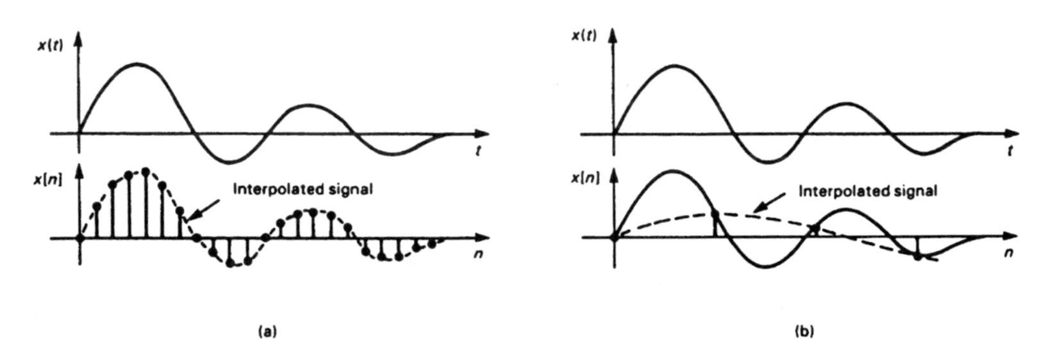

Fig. 1.7 Interpolation of samples. (a) Sampling rate sufficiently high; (b) loss of signal information due to a low sampling rate.

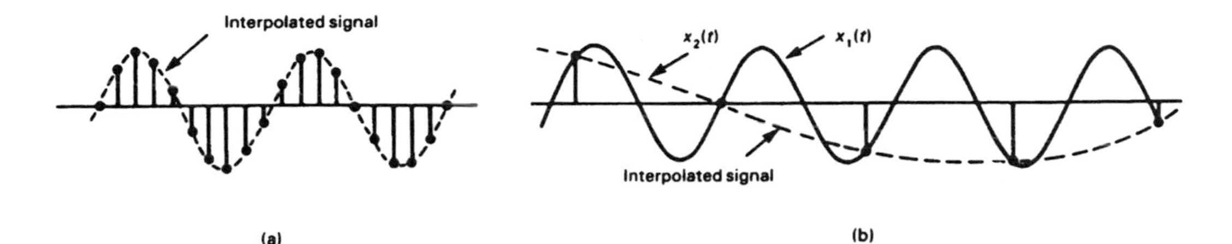

Fig. 1.8 Samples of a sinusoidal signal. (a) Adequate sampling; (b) low sampling rate results in an aliasing error.

which always suggests a signal having a lower frequency content than the original.

This loss of high-frequency information is illustrated further in Figure 1.8 with the samples of a sinusoidal signal. In Figure 1.8a the signal has been sampled adequately so that its waveform can be recovered by interpolation. However, the sinusoid shown in Figure 1.8b has been sampled at a relatively low rate and we see that interpolation gives not the original signal $x_1(t)$, but a sinusoid $x_2(t)$ at a lower frequency. This kind of confusion in the interpretation of samples is generally known as an *aliasing error*, and signals sharing the same set of samples are known as *aliased signals*.

Once we have lost information by inadequate sampling, we are unable to recover the waveform of the original signal by interpolation or by any other means.

This applies to all types of signal. For example, the signal $x(t)$ and the interpolated signal shown in Fig. 1.7b are aliased signals.

Worked Example 1.8

Show that the following are aliased signals: $a(t) = \cos \omega_0 t$ and $b(t) = \cos(\omega_0 + \omega_s)t$, where $\omega_s = 2\pi f_s$ is the sampling frequency in rad s^{-1}.

Solution: On sampling the signals we find the sample values

$$a[n] = a(nT) = \cos n\omega_0 T \text{ and } b[n] = b(nT) = \cos n(\omega_0 + \omega_s)T.$$

Now, $\omega_s = 2\pi f_s = 2\pi/T$, therefore $b[n] = \cos(n\omega_0 T + 2n\pi)$

If we now use the general result: $\cos(\alpha + 2n\pi) = \cos \alpha$, we see that the two signals have the same samples: $a[n] = b[n] = \cos n\omega_0 T$. The two sinusoids differing in frequency by the sampling frequency are thus aliased signals.

It is not difficult to extend the results of Worked Example 1.8 and show that sinusoidal signals differing in frequency by multiples of the sampling frequency f_s will have the same sample values. This means that if a signal description includes sinusoidal components at frequencies $f_0 + f_s$, $f_0 + 2f_s$ and so on, then their samples will be indistinguishable from those of a component at the lower frequency, f_0. The result is an incorrect interpretation when we interpolate the signal samples to recover its waveform.

If a signal is *bandlimited* however, containing components up to a maximum frequency f_b, then we can always avoid errors due to aliasing by sampling in accordance with the *sampling theorem* which states:

> A continuous signal with frequency components in the range $f = 0$ to $f = f_b$ Hz can be reconstructed from a sequence of equally spaced samples, provided that the sampling frequency exceeds $2f_b$ samples/second.

This means in effect that the highest frequency component of a signal must be sampled more than twice per cycle if the original signal is to be recovered by interpolation. Thus, if we have a signal occupying the frequency range $f = 0$ to $f = 10\,\text{kHz}$, the minimum sampling rate for a complete representation of the signal must exceed 20 000 samples/second.

It is usual to assume for practical purposes that all the significant components of a signal lie within its bandwidth and that components at greater than the bandwidth frequency have negligible values. The question of frequency content is so important that it is usual practice to transmit a signal through a low-pass filter prior to sampling. This filter has a sharp cut-off at a well-defined frequency and for obvious seasons is known as an *anti-aliasing filter*. To be effective the filter cut-off frequency must be less than half the sampling frequency and in critical applications is very often less than $f_s/5$.

The implications of the sampling theorem will be considered in greater detail in Chapter 7 when we look more closely at the relationship between a signal $x(t)$ and its sample sequence $x[n]$. Meanwhile, we shall treat the sampling process quite informally and use it as a convenient means of representing the values of real experimental signals.

Summary

In this chapter we have introduced a range of notation and symbols to enable us to represent signals and systems in both the continuous- and discrete-time domains. This is essential if we are to take account of modern electronic systems which often handle signal data in the form of numerical sequences. In practice, discrete-time signals are often obtained by taking samples from a continuous-time analogue signal.

The most elementary representation of a system is a 'black box' in which input and output are related by a set of mathematical equations known as the system model. The model serves to define the function of a system but otherwise gives no information about its physical content. In the case of discrete-time systems the model can often be drawn as a system diagram showing a set of interconnected summation blocks, multipliers and delay elements.

Perhaps the most important assumption that we can make about a system is that it is linear with properties that do not vary with time. We have set out the rules governing the behaviour of such systems and highlighted some of the difficulties that can arise when we compare behaviour of 'ideal' models with practical devices.

Linear systems obey the principle of superposition. Thus, given a signal expressed as a sum of elementary components, we can find the response of a system by adding together the individual component responses. We have reviewed a number of different signal categories in preparation for later work on signal analysis and defined a number of important signal parameters. An important conclusion has been that signals can be divided broadly according to whether they have finite energy or finite power and it has been suggested that these types of signal are treated rather differently for the purposes of analysis.

We have revised the properties of odd, even and orthogonal signals with a view to handling certain types of integral that will occur repeatedly in the following chapters. The main lesson here is that signal analysis is concerned as much with shape and symmetry as it is with solving difficult integrals.

The final section introduced the sampling theorem. We saw that if the waveform of a signal is to be reconstructed from its samples, the signal should be bandlimited to a maximum frequency f_B and the sampling frequency chosen so that $f_s > 2f_B$. If this condition is not met we lose information about the rapidly changing parts of a signal.

Problems

1.1 Causal signals $a[n] = n$ and $b[n] = 2^n$ are applied to a summation block. Write down the values of the first few terms in the output sequence.

1.2 The unit-step sequence is defined to be the causal sequence $u[n] = 1$ for $n \geqslant 0$. Plot the sequences $2u[n]$, $u[n-3]$, $u[n] + u[n-4]$ and $u[n] - u[n-5]$.

1.3 Samples are taken from an analogue signal in the range $t = 0$ to $t = NT$, where T is the sampling interval. Show that the area under the signal graph can be approximated by the summation

$$A \simeq T \sum_{n=0}^{N-1} x[n].$$

1.4 In view of problem 1.3, suggest how the integral of a signal from $t = 0$ to $t = NT$ could be approximated using a discrete-time integrator and multiplier block.

1.5 An energy signal $x(t)$ is applied to a resistive load. Show that the signal power averages to zero over a very long measurement interval.

1.6 A sinusoidal signal $x(t) = A\sin(\omega t + \theta)$ can be expressed as the sum of orthogonal components:

$$x(t) = A\cos\theta \sin\omega t + A\sin\theta \cos\omega t.$$

Show that the average power of the signal is equal to the sum of the component powers.

1.7 Verify that the following pairs of harmonic sinusoids are orthogonal over any interval of length $T_0 = 2\pi/\omega_0$. (a) $\cos m\omega_0 t$ and $\sin n\omega_0 t$, (b) $\cos m\omega_0 t$ and $\cos n\omega_0 t$, $m \neq n$, and (c) $\sin m\omega_0 t$ and $\sin n\omega_0 t$, $m \neq n$, where m and n are integers.

1.8 The sinusoidal signal in Worked Example 1.1 has a frequency $f_0 = 1\,\text{Hz}$ and is sampled at the frequency $f_s = 8\,\text{Hz}$. The fundamental period of the resulting discrete-time sequence $x[n]$ is $N = 8$.

Calculate and sketch the sequence $x[n]$ for a) $f_0 = 2\,\text{Hz}$ and b) $f_0 = 3\,\text{Hz}$, with $f_s = 8\,\text{Hz}$ as before.

What will be the fundamental period of $x[n]$ in each case?

1.9 If a sinusoid of frequency f_0 is sampled at the instants $t = nT$, $n = 0, \pm 1, \pm 2, \ldots$ the sample sequence can be written as

$$x[n] = \sin(2\pi f_0 nT) = \sin\left(2\pi \frac{f_0}{f_s} n\right).$$

Show that if $x[n]$ is to be strictly periodic with *integer* period N then f_0/f_s must be a rational number, that is:

$$\frac{f_0}{f_s} = \frac{k}{N}$$

where k is an integer. The fundamental period is the *smallest* value of N which satisfies this relationship.

1.10 Samples are to be taken from a record of a continuous-time signal of duration 100 ms. The signal contains sinusoidal components with frequencies up to 250 Hz. Choose from the following options the number of samples that would be *sufficient* to give a complete representation of the signal:

a) 20; b) 40; c) 60; d) 80; e) 100.

Time-domain Models 2

Objectives

- ☐ To introduce the unit-sample sequence and the unit impulse function as basic components in modelling continuous- and discrete-time signals.
- ☐ To introduce the unit-impulse response function as the characteristic time-domain response of a linear system.
- ☐ To explain how the process of convolution relates the response of a linear system to the input signal and the system's impulse response.
- ☐ To give a physical interpretation of the unit-impulse function and the impulse response for a continuous system.

In this chapter we shall be concerned with the representation of signals, and the input-output behaviour of linear time-invariant systems, in the time domain. There are two main aims. The first is to establish an approach to modelling an arbitrarily complex continuous- or discrete-time signal by representing the signal as a sum of simpler components. Given such a model our second aim is to find the output of a system from a knowledge of the input signal and the overall system behaviour. We shall begin by looking at some simple models of discrete-time processors.

The assumption of linearity, and the principle of superposition in particular, is of great importance in modelling signals and systems and it underpins all our work in this chapter. The properties of linear time-invariant systems were discussed in Chapter 1.

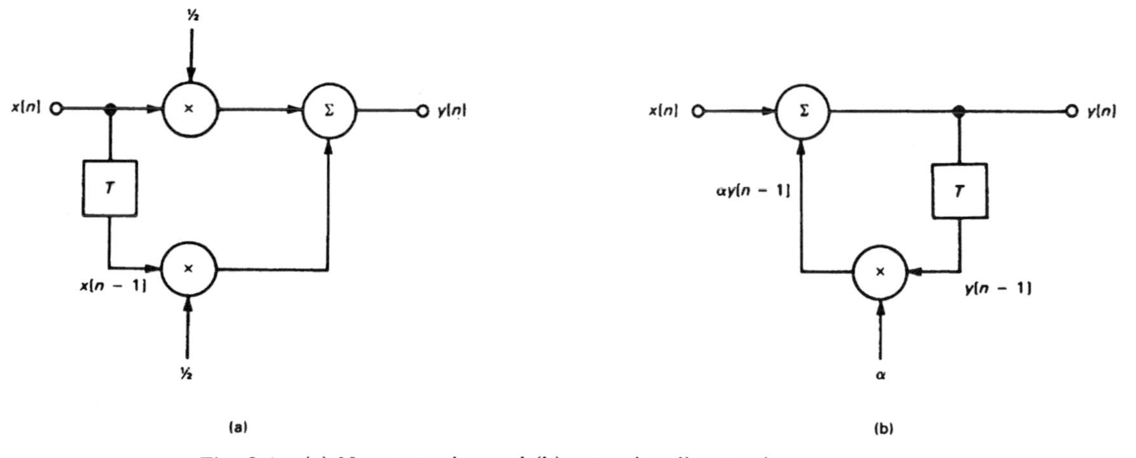

Fig. 2.1 (a) Non-recursive and (b) recursive discrete-time processors.

Discrete-time systems

Figure 2.1 shows block diagrams of two discrete-time signal processors and demonstrates the use of the symbols introduced in Chapter 1 to represent the operations of summation, multiplication and delay. Figure 2.1a represents the processor defined by the expression

The operations performed by a discrete-time processor can be implemented either as a set of instructions on a digital computer, or by the use of special-purpose electronic circuits. In either case we represent the overall signal-processing operation of a discrete-time system as an interconnection of summation, multiplication and time-delay blocks.

A sequence $x[n]$ is often the result of periodically sampling a continuous-time signal. It is often useful to think of the output sequence $y[n]$ as a sampled version of the processed continuous signal.

A recursive realization always involves feedback.

$$y[n] = \frac{x[n] + x[n-1]}{2}$$
$$= \tfrac{1}{2}x[n] + \tfrac{1}{2}x[n-1]. \tag{2.1}$$

In this example, the output $y[n]$ is equal to the sum of the input $x[n]$ and the delayed input $x[n - 1]$, each scaled by a factor of $\frac{1}{2}$. A single delay block is used to relate the delayed sequence $x[n - 1]$ to the input sequence $x[n]$. Multiplier blocks are then used to scale each sequence before they are combined by the summing block to form the output $y[n]$.

In the system shown in Figure 2.1b the output sequence depends not only on the input but also on previous values of the output. The processing operation is defined by the expression

$$y[n] = x[n] + \alpha y[n-1]. \tag{2.2}$$

In this case a single delay block is used to relate the previous output $y[n - 1]$ to the present output $y[n]$. The delayed output $y[n - 1]$ is then scaled by a factor of α by the multiplier before being added to the input sequence. The output of the summing block is the output $y[n]$ of the system. Because the output involves not only the input but also the past value $y[n - 1]$ of the output, the system is called *recursive*. By contrast, the system shown in Figure 2.1a is called *non-recursive*, because its output is expressed solely in terms of the input sequence $x[n]$.

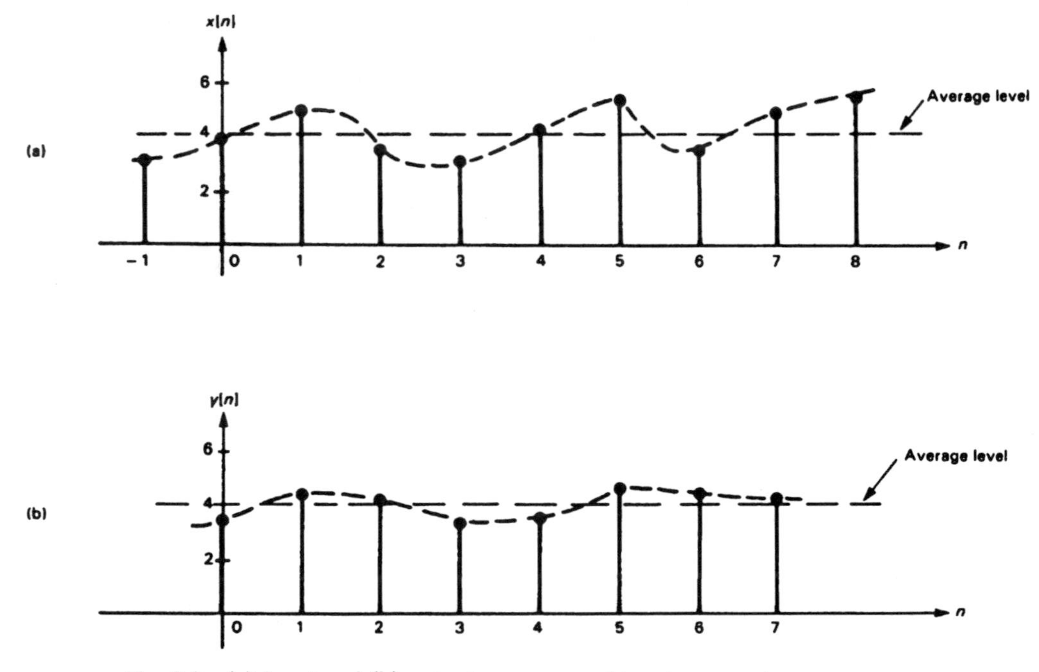

Fig. 2.2 (a) Input and (b) output sequences of two-term moving averager.

Table 2.1

n	0	1	2	3	4	5	6	7
$x[n]$	4.0	5.0	3.6	3.1	4.2	5.5	3.7	5.0
$x[n-1]$	3.2	4.0	5.0	3.6	3.1	4.2	5.5	3.7
$y[n] = (x[n] + x[n-1])/2$	3.6	4.5	4.3	3.4	3.7	4.9	4.6	4.4

Both of these simple processors can act as low-pass filters, by reducing the amplitude of rapidly varying components in the input sequence. We can demonstrate this effect by a simple example. Figure 2.2a shows a sequence $x[n]$ such as may be obtained from periodically sampling a continuously varying measurement signal. Such a signal may represent, for example, the level of fuel in the tank of a moving vehicle. The motion of the vehicle causes the fuel to slosh in the tank, and hence the signal level varies. We wish to process this sampled signal in order to reduce the signal fluctuations and get a better estimate of the average level of fuel in the tank at any time.

A simple way to produce a smoother sequence of values is to take the average of successive pairs of input values. This is exactly what happens if we apply the sequence to the non-recursive processor shown in Figure 2.1a. The processor effectively adds each input value $x[n]$ to the previous input value $x[n-1]$ and divides by two. Each output value $y[n]$ is, therefore, the average of the current and previous inputs. Table 2.1 shows the result of this processing operation on the input sequence in Figure 2.2a.

The table contains the input sequence $x[n]$ and the delayed sequence $x[n-1]$. Each term in the output sequence $y[n]$ of the processor is the average of $x[n]$ and $x[n-1]$ and represents an estimate, at any instant n, of the average value of the level of fuel in the tank. Because the averaging operation effectively moves along the input sequence the processor is called a two-term *moving averager*. The smoothing effect of the moving averager is illustrated clearly in Figure 2.2b. Thus, by comparing input and output, we see that the amplitude of the fluctuations in the input sequence has been considerably reduced.

Rapid changes in a signal indicate the presence of higher frequency components. We will see later how to analyse a signal to find out about the distribution of these components. Any processor which attenuates higher-frequency components. We will affecting slower, long-term changes in the signal can be thought of as a form of low-pass filter.

It may be shown that the recursive processor in Figure 2.1b also provides a smoothing, or low-pass filtering, effect on an input sequence. We shall be looking in more detail at the filtering effect of discrete-time systems in Chapter 5.

The recursive and non-recursive processors in Figure 2.1 are simple examples of discrete-time systems. In the general case, the current values of the output may depend on the N and M past values of the input and output. Thus the general form of the system equation is

$$\begin{aligned} y[n] = {} & a_0x[n] + a_1x[n-1] + a_2x[n-2] + \dots + a_Nx[n-N] \\ & - b_1y[n-1] - b_2y[n-2] + \dots - b_My[n-\mathrm{M}] \end{aligned} \quad (2.3)$$

or in more compact summation notation:

$$y[n] = \sum_{i=0}^{N} a_i x[n-i] - \sum_{k=1}^{M} b_k y[n-k]. \quad (2.4)$$

The terms 'recursion equation' and 'recurrence formula' are also used.

The equation describing the operations performed by a discrete-time system is called a *linear difference equation*, or *recurrence equation.*

Worked Example 2.1

Draw the block diagram of the discrete-time system represented by the recurrence equation

$$y[n] = x[n] + 0.2x[n-1] + 0.5y[n-1].$$

Solution: There are many equivalent ways of drawing the block diagram of a given system. Figure 2.3 shows an approach that may be used in the present Example.

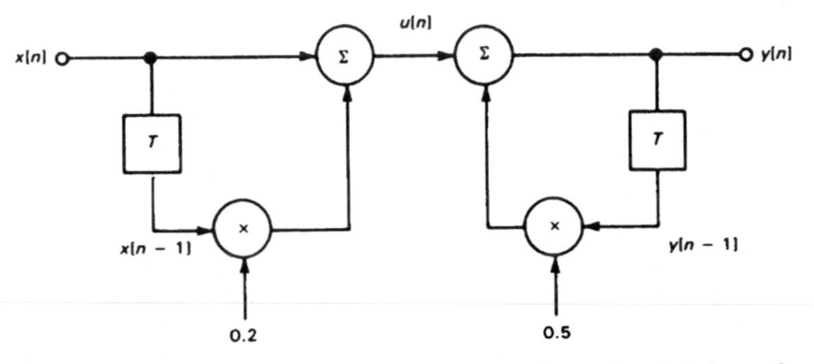

Fig. 2.3 Block diagram of $y[n] = x[n] + 0.2x[n-1] + 0.5y[n-1]$.

The input terms $x[n]$ and $0.2x[n-1]$ are combined to form an intermediate sequence $u[n]$, where

$$u[n] = x[n] + 0.2x[n-1]$$

so that the output sequence can be written

$$y[n] = u[n] + 0.5y[n-1].$$

The sequence $0.5y[n-1]$ is obtained by delaying and scaling the output $y[n]$. The delayed output sequence is then added to $u[n]$ to form the overall input-output relationship.

Exercise 2.1

Draw the block diagrams of the processors described by the equations:

(i) $y[n] = \dfrac{x[n] + x[n-1] + x[n-2]}{3}$

(ii) $y[n] = x[n] + y[n-1] - 0.5y[n-2]$.

Unit-sample response and convolution

One way of investigating the behaviour of a system is to observe its response to an input test signal. Of course we shall not be carrying out tests on a physical system but rather using the idea to develop a mathematical technique which will allow us to predict the response of the system to any input signal. Although we are free to choose any signal we like as a test input, we base our choice usually on two main requirements. Firstly, the test signal model must be mathematically simple to handle. Secondly, the test signal must form a basic component from which we can build up more complex signal models in a straightforward manner.

If we express our complex signal models in terms of a sum of simpler components then, assuming that we are dealing with linear time-invariant system models, we can use the principle of superposition to work out the response of a system to an arbitrary input signal. We look first at how we can model discrete-time signals in terms of sums of simple sequences.

In discrete-time signal models the *unit-sample sequence*, or *delta sequence* is the basic building block. The unit-sample sequence is represented by the symbol $\delta[n]$ and is defined as the sequence

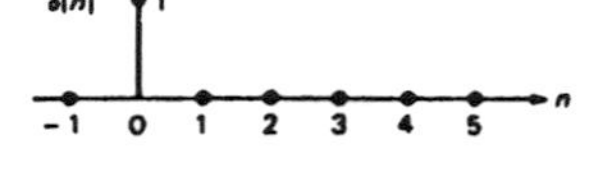

$$\delta[n] = 1, 0, 0, 0, 0, \ldots \tag{2.5}$$

that is, the value of $\delta[n]$ is equal to 1 only at $n = 0$. For any other value of n, $\delta[n]$ is zero.

A unit-sample sequence that has been delayed for k sample periods is written as $\delta[n-k]$. The value of $\delta[n-k]$ is equal to 1 only for $n = k$. At any other value n, $\delta[n-k]$ is zero. As an example the sequence $\delta[n-3]$ is defined as

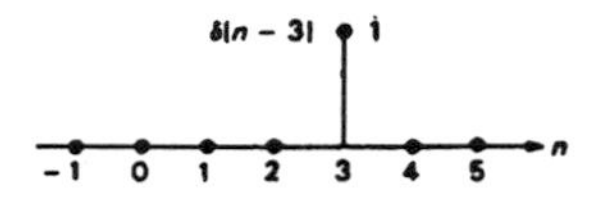

$$\delta[n-3] = 0, 0, 0, 1, 0, 0, 0, \ldots$$

The unit-sample sequence may be both delayed and scaled by a constant multiplier. So, for example, the sequence $1.5\delta[n-2]$ is

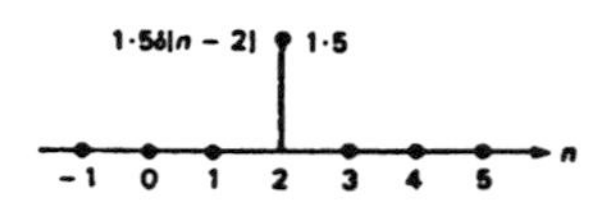

$$1.5\delta[n-2] = 0, 0, 1.5, 0, 0, 0, \ldots$$

We can now represent any discrete-time signal as a sum of scaled, delayed unit-sample sequences. The sequence shown in Figure 2.4 overleaf

$$x[n] = 3, 1, 2, -1, 0, 0, \ldots$$

can be written as the sum of four sequences

$$x[n] = (3, 0, 0, 0, 0, \ldots) + (0, 1, 0, 0, 0, \ldots) + (0, 0, 2, 0, 0, \ldots) + (0, 0, 0, -1, 0, \ldots)$$

Each sequence is a scaled, delayed unit-sample sequence so we can write $x[n]$ as the sum

$$x[n] = 3\delta[n] + \delta[n-1] + 2\delta[n-2] - \delta[n-3].$$

In general, we can express any sequence as the sum of scaled, delayed unit-sample sequences. Hence an arbitrary sequence of the form

$$x[n] = \ldots x[-2], x[-1], x[0], x[1], x[2], x[3], \ldots$$

can be written as the sum

$$\begin{aligned} x[n] &= \ldots x[-2]\,\delta[n+2] + x[-1]\,\delta[n+1] + x[0]\,\delta[n] \\ &\quad + x[1]\,\delta[n-1] + x[2]\,\delta[n-2]\ldots \\ &= \sum_{k=-\infty}^{\infty} x[k]\,\delta[n-k]. \end{aligned} \tag{2.6}$$

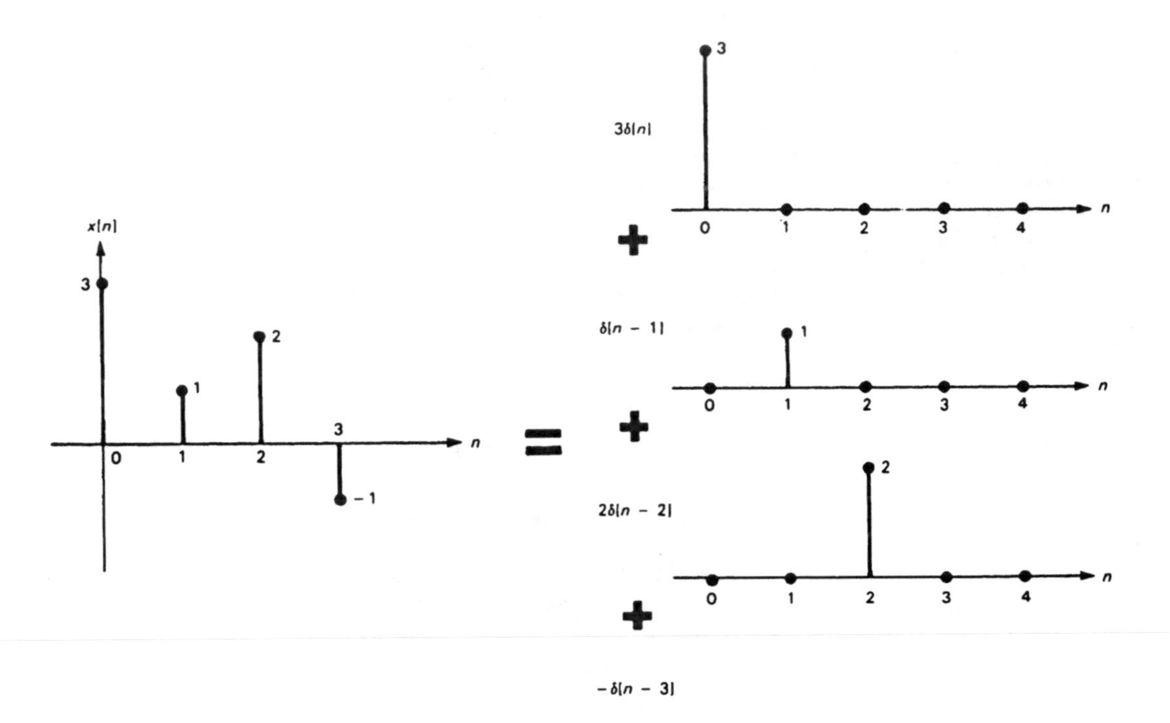

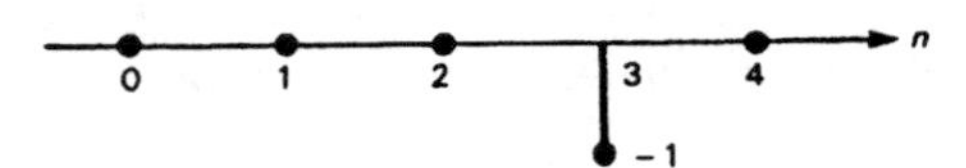

Fig. 2.4 Sequence expressed as a sum of scaled, delayed unit-sample sequences.

The unit-sample response

Suppose now that the unit-sample sequence $\delta[n]$ forms the input to a discrete-time processor and, in response, the processor produces the output sequence $h[n]$:

$$\delta[n] \rightarrow h[n]. \tag{2.7}$$

The sequence $h[n]$ is called the *unit-sample response* or *impulse response* of the processor.

Worked Example 2.2 A discrete-time processing operation is defined by the recurrence equation

$$y[n] = x[n] + 2x[n-1] + 3x[n-3].$$

What is the unit-sample response $h[n]$ of the processor?

Solution: Writing the equation of the processor out in full gives

$$y[n] = x[n] + 2x[n-1] + 0x[n-2] + 3x[n-3].$$

When the input $x[n]$ is the unit-sample sequence $\delta[n]$ the output $y[n]$ is, by definition, the unit-sample response $h[n]$. Replacing $y[n]$ by $h[n]$, $x[n]$ by $\delta[n]$, $x[n-1]$ by $\delta[n-1]$, and so on in the recurrence equation gives

$$h[n] = \delta[n] + 2\delta[n-1] + 0\delta[n-2] + 3\delta[n-3].$$

In other words the unit sample-response $h[n]$ is simply a sequence containing four terms. Each term in $h[n]$ is equal to the corresponding multiplier coefficient in the recurrence equation.

Notice that there is no contribution to the output from the term $x[n-2]$. Hence the corresponding term $h[2]$ in the unit-sample response will be zero. The sequence $h[n]$ is given directly by the multiplier coefficients:

$$h[n] = 1, 2, 0, 3, 0, 0, \ldots$$

where all undefined coefficients are taken to be zero.

The form of the unit-sample response provides a useful way of classifying a linear discrete-time processor. Consider first a non-recursive system whose output $y[n]$ depends only on the present and past values of the input $x[n]$. If the output contains contributions from up to N past inputs, the recurrence equation contains $N+1$ terms:

$$y[n] = \sum_{i=0}^{N} a_i x[n-i]$$

$$= a_0 x[n] + a_1 x[n-1] + a_2 x[n-2] + \ldots + a_N x[n-N] \quad (2.8)$$

The unit-sample response $h[n]$ of this system is therefore a sequence containing $N+1$ terms. Each term in the unit-sample response sequence is equal to the corresponding multiplier coefficient a_i in the recurrence equation. Hence

$$h[n] = a_0, a_1, a_2, \ldots a_{N-1}, a_N, 0, 0, 0, \ldots$$

A practical non-recursive processor will contain a finite number N of delay elements. So the corresponding unit-sample response will contain a finite number $(N+1)$ of terms. Such a processor is referred to as a *finite-impulse response* or FIR system.

The moving averager discussed earlier is a simple example of an FIR system. Recall that the recurrence equation of the processor was

$$y[n] = \tfrac{1}{2}x[n] + \tfrac{1}{2}x[n-1].$$

We can write down the unit-sample response of the moving averager directly from the equation, giving

Using the delta sequence notation we can also write this result in the form

$h[n] = \delta[n]/2 + \delta[n-1]/2$

$$h[n] = \tfrac{1}{2}, \tfrac{1}{2}, 0, 0, \ldots$$

The sequence $h[n]$ is given directly by the multiplier coefficients in the recurrence equation.

Consider now the recursive processor defined by the equation

$$y[n] = x[n] + \alpha y[n-1]$$

In this case the current output is related to both the current input and the previous output. As Figure 2.1b showed, we can represent this system by incorporating a single delay element in a feedback path. The presence of the feedback has an important effect on the unit-sample behaviour of the system. Unlike the FIR case we cannot write down the unit-sample response simply by looking at the coefficients in the processor equation. What we must do is to calculate the sequence $h[n]$ by following the order of operations defined by the equation. For this example we shall take the feedback coefficient α as $\frac{1}{2}$.

Whenever we work with equations of systems containing delays in feedback paths we find that values have to be carried over from one calculation to the next. One way to keep track of these calculations is to set up a table of values.

Table 2.2

n	0	1	2	3	4	5	6	
$x[n]$	1	0	0	0	0	0	0	
$y[n-1]$	0	1	1/2	1/4	1/8	1/16	1/32	...
$y[n] = x[n] + \frac{1}{2}y[n-1]$	1	1/2	1/4	1/8	1/16	1/32	1/64	

Do not assume that a recursive realization always gives an infinite impulse response. See, for example, end of chapter problem 2.8.

The first row in the table gives the value of n. The table then has one row for each term in the processor equation. Since the input is the unit-sample sequence we can immediately fill in the row for $x[n]$ with the entries 1, 0, 0, 0, 0, ... In the first column, where $n = 0$, the entry for $y[n-1]$ is zero, because we assume that there is no output from the system before the input is applied. We work out the output sample $y[0]$ by combining the values in the first column in the way defined by the processor equation. The result $y[0] = 1$ is entered under $n = 0$ in the last row.

We now move to the second column, where $n = 1$, and transfer the value of $y[0]$ to the $y[n-1]$ row, as shown by the arrow. As before we work out the value for $y[1]$ by combining terms in the $n = 1$ column to give $y[1] = \frac{1}{2}$.

The arrows in the table show how we continue this process, by carrying terms forward from column to column, each output value contributing to future outputs. In this example the values in the output sequence are getting smaller as n increases. Unlike the FIR case, however, they will never reach zero (and remain at zero) in a finite number of terms. The output sequence is infinitely long and defines the processor as an *infinite impulse response* or *IIR* system.

Exercise 2.2 By constructing a table, as shown above, find the unit-sample response of the IIR system defined by the equation

$$y[n] = x[n] + y[n-1] - \tfrac{1}{2}y[n-2].$$

Assume that the output of the processor is zero before the input is applied.

The convolution–sum relationship

The unit-sample response is a fundamental characteristic of a discrete-time linear system. Knowing the form of the unit-sample response of a given system allows us to work out, in principle at least, the response of the system to any arbitrary input. We can show this by using the basic properties of homogeneity, time-invariance and superposition.

Firstly, the property of homogeneity tells us that if $h[n]$ is the unit-sample response then the response to the scaled unit-sample sequence $a\delta[n]$ is simply the scaled sequence $ah[n]$, hence we have

$$a\delta[n] \rightarrow ah[n]. \tag{2.9}$$

Secondly, the property of time-invariance means that a scaled and delayed unit-sample input sequence will produce a scaled and delayed response, hence

$$a\delta[n-k] \rightarrow ah[n-k]. \tag{2.10}$$

Finally, the principle of superposition means that if the input to a system is made up of the sum of individual components then the overall response of the system can be thought of as the sum of the responses to the individual components. So if, for example, an input sequence is represented as the sum

$$x[n] = 3\delta[n] + \delta[n-1] + 2\delta[n-2] - \delta[n-3]$$

then each component sequence will produce a scaled, delayed unit-sample response:

$$\begin{array}{llll} 3\delta[n] & \rightarrow 3h[n] & 2\delta[n-2] \rightarrow & 2h[n-2] \\ \delta[n-1] \rightarrow & h[n-1] & -\delta[n-3] \rightarrow & -h[n-3] \end{array} \tag{2.11}$$

In this case the overall response $y[n]$ to the input sequence $x[n]$ will be the sum of the scaled, delayed unit-sample responses:

$$y[n] = 3h[n] + h[n-1] + 2h[n-2] - h[n-3].$$

Worked Example 2.3

The input sequence

$$x[n] = 3, 1, 2, -1, 0, 0, \ldots$$

is applied to a discrete-time processor with the unit-sample response

$$h[n] = 1, 2, 1, 0, 0, \ldots$$

What is the resulting output sequence of the processor?

Solution: Figure 2.5 shows the input sequence as a sum of scaled and delayed unit-sample sequences. The right-hand side of the figure shows the response of the processor to each input sequence, where

$$\begin{aligned} 3\delta[n] \rightarrow 3h[n] &= 3, 6, 3, \;\; 0, \;\; 0, \;\; 0, 0, \ldots \\ \delta[n-1] \rightarrow h[n-1] &= 0, 1, 2, \;\; 1, \;\; 0, \;\; 0, 0, \ldots \\ 2\delta[n-2] \rightarrow 2h[n-2] &= 0, 0, 2, \;\; 4, \;\; 2, \;\; 0, 0, \ldots \\ -\delta[n-3] \rightarrow -h[n-3] &= 0, 0, 0, -1, -2, -1, 0, \ldots \end{aligned}$$

The overall output sequence $y[n]$ is the sum of the individual responses, so

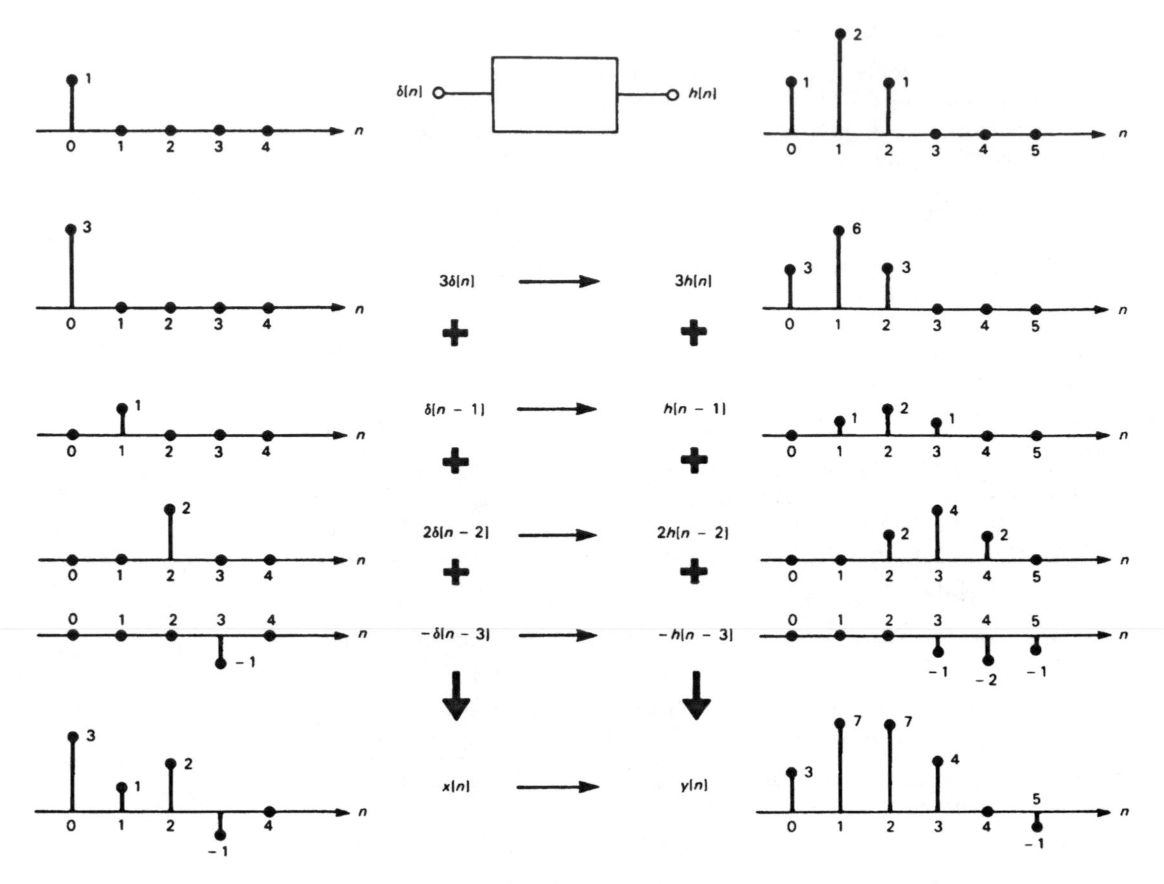

Fig. 2.5 Principle of discrete-time convolution.

$$
\begin{aligned}
y[0] &= 3 + 0 + 0 + 0 = 3 \\
y[1] &= 6 + 1 + 0 + 0 = 7 \\
y[2] &= 3 + 2 + 2 + 0 = 7 \\
y[3] &= 0 + 1 + 4 - 1 = 4 \\
y[4] &= 0 + 0 + 2 - 2 = 0 \\
y[5] &= 0 + 0 + 0 - 1 = -1
\end{aligned}
$$

All values of $y[n]$ for $n > 5$ are zero. Hence $y[n]$ is

$$y[n] = 3, 7, 7, 4, 0, -1, 0, 0, \ldots$$

We can generalize this result to allow us to handle any input sequence. We know from Equation 2.6 that an arbitrary sequence can be expressed as the sum

$$x[n] = \sum_{k=-\infty}^{\infty} x[k]\,\delta[n - k].$$

Each component sequence will produce a response

$$x[k]\delta[n-k] \rightarrow x[k]h[n-k]$$

so the overall response will be the sum of the individual responses

$$y[n] = \sum_{k=-\infty}^{\infty} x[k]h[n-k]. \qquad (2.12)$$

Given the input sequence $x[n]$, therefore, and knowing the unit-sample response of the system $h[n]$, this expression allows us to work out the resulting output sequence $y[n]$. This process is called *convolution* and the expression given above is sometimes called the *convolution sum*. In discrete-time systems convolution is an operation performed on sequences. The convolution operation is usually denoted by the * symbol, hence we can write

$$y[n] = x[n] * h[n] \qquad (2.13)$$

and we say that the resulting sequence $y[n]$ is the *convolution of* $x[n]$ with $h[n]$. The order in which two sequences are convolved is unimportant. In other words we can write

$$y[n] = \sum_{k=-\infty}^{\infty} x[k]h[n-k] = \sum_{k=-\infty}^{\infty} h[k]x[n-k]$$
$$= x[n] * h[n] \qquad = h[n] * x[n] \qquad (2.14)$$

We interpret this result to mean that a linear time-invariant system with an input $x[n]$ and a unit-sample response $h[n]$ will produce the same output $y[n]$ as a system with an input $h[n]$ and a unit-sample response $x[n]$.

The result that the output $y[n]$ is related to $x[n]$ and $h[n]$ by the operation of convolution is a consequence *only* of our initial assumptions that the system is linear and time-invariant. It is a *terminal property* and does not depend on the detailed structure of the system. The result is significant because it demonstrates that, in principle, a single measurement of the unit-impulse response of a system contains sufficient information to allow us to predict completely the response of the system to any input.

Because the order is unimportant to the result the operation of convolution is said to be *commutative*.

We cannot differentiate between the characteristics of the input signal and the characteristics of the system by looking only at the output.

Exercise 2.3

$a[n]$ and $b[n]$ are two sequences defined by

$$a[n] = 2, 3, 1 \text{ and } b[n] = 1, 2, 2, 1$$

Work out the output sequence $y[n]$ that would result if $a[n]$ and $b[n]$ were respectively the input and the unit-sample response of a system. Show that the output sequence would be unchanged if $b[n]$ was the input and $a[n]$ was the unit-sample response.

Why does the convolution of a three-term sequence with a four-term sequence produce a six-term sequence?

In principle the convolution of input and impulse-response sequences is a straight-forward operation which can be performed easily by a computer. The impulse responses of IIR systems, however, contain an infinite number of terms and in practice we can deal only with finite-length sequences. In such cases, therefore, the inpulse-response sequence must be truncated, according to some criterion, before convolution can be carried out.

Discussion of the effects of the different methods of sequence truncation, called 'windowing', is beyond the scope of this book. A detailed treatment of windowing is to be found in Oppenheim, A.V. and Schafer, R.W., *Discrete-Time Signal Processing*, Prentice-Hall, 1989. Other practical aspects of digital processing are discussed in Lynn, P. and Fuerst, W. *Introductory Digital Signal Processing*, John Wiley, 1989.

Convolution for continuous systems

In the discrete-time case we used the unit-sample sequence $\delta[n]$ as a basic element in modelling an arbitrary sequence. Looking at the overall response of a system as the sum of responses to scaled, time-shifted unit-sample sequences

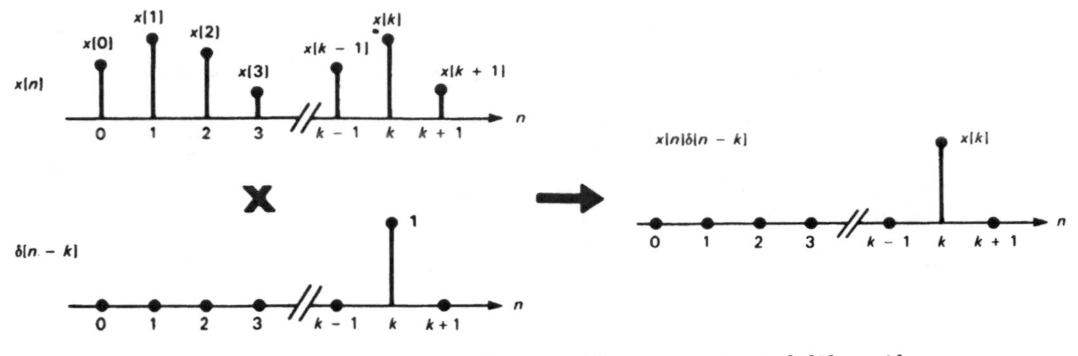

Fig. 2.6 Sifting property of $x[n]\delta[n-k]$.

led us in a straightforward manner to the expression for the discrete convolution sum. In this section we are going to proceed in a similar way to develop an analogous convolution relationship for continuous-time systems. We start by considering an important property of the unit-sample sequence.

Suppose that we multiply an arbitrary sequence $x[n]$ by the unit sample sequence $\delta[n-k]$. As Figure 2.6 shows, the product $x[n]\delta[n-k]$ is a sequence containing all zero values except for the value $x[k]$ at $n = k$. Multiplying by $\delta[n-k]$ has acted rather like a sieve; all values of $x[n]$ have been suppressed except $x[k]$. Now if we add up the terms in the sequence $x[n]\delta[n-k]$ we get

$$\sum_{n=-\infty}^{\infty} x[n]\delta[n-k] = \ldots + 0 + 0 + 0 + x[k] + 0 + 0 + 0 + \ldots$$

$$= x[k] \tag{2.15}$$

Summing the product $x[n]\delta[n-k]$ over all n, therefore, gives a *number* which is equal to the value of the sequence $x[n]$ at $n = k$. Effectively the procedure has selected or 'sifted' out of $x[n]$ the value of the sample at $n = k$.

Let us now consider how we might apply this idea of a continuous signal. Let us imagine that there is a time-domain function $\delta(t)$ which plays a role in the continuous case similar to that played by $\delta[n]$ in the discrete case. We are not yet saying what sort of function $\delta(t)$ is, only that its sifting property is analogous to that of $\delta[n]$.

Using the notation $\delta(t-\tau)$ to indicate a time-shifted version of $\delta(t)$ we can proceed by analogy with the discrete case to select or sift out the value $x(\tau)$ of a continuous time signal $x(t)$ at a time $t = \tau$. As before we form the product $x(t)\delta(t-\tau)$, but because we are now dealing with continuous signals we use an integral to represent a continuous summation over all time. The resulting *sifting* or *sampling* integral

$$\int_{-\infty}^{\infty} x(t)\,\delta(t-\tau)\,\mathrm{d}t = x(\tau) \tag{2.16}$$

The sifting or sampling property of $\delta(t)$

picks out the value of $x(t)$ at the instant $t = \tau$, that is $x(\tau)$. In fact, this integral is used as a *definition* of the fundamental sifting property of $\delta(t)$. In other words, we have defined $\delta(t)$ in terms of what it *does*, rather than what it *is*. $\delta(t)$

or its time-shifted version $\delta(t - \tau)$ is called the *impulse function*, or the (Dirac) *delta function*.

One consequence of the integral definition of Equation 2.16 is that $\delta(t - \tau)$ has the dimensions 1/[Time]. For example, if $x(t)$ represents a time-varying voltage then the spot value $x(\tau)$ must also have the dimensions of voltage. If $\delta(t - \tau)$ has the dimensions 1/[Time] then the product $x(t)\,\delta(t - \tau)\,\mathrm{d}t$ in the sifting integral will have the dimensions of voltage, as required.

Paul Adrian Maurice Dirac won the 1933 Nobel Prize for Physics which he shared with Erwin Schrödinger. He used the delta function in his work on quantum mechanics and developed mathematical techniques for handling the function.

Interpretation of the delta function and the impulse response

Unlike the unit-sample sequence in the discrete-time case, the impulse, or Dirac delta, function $\delta(t)$ is a rather sophisticated mathematical concept. Indeed $\delta(t)$ is not a function at all in the ordinary sense. As we have seen in the previous section we define $\delta(t)$ in terms of what it does and not what it is. Although this might seem to be a strange way to proceed it turns out that the delta function can be used to great effect to simplify the description of signals. The delta function is not a model of any particular signal but rather a mathematical entity which we can learn to interpret physically.

Although the delta function was used by Dirac in the late 1920s it was not until the early 1950s when L. Schwartz in his *Théorie des Distributions*, vols 1 and 2, Hermann et Cie, Paris, 1950 and 1951, put the function on a mathematically rigorous basis. A detailed treatment of the delta function in terms of the theory of generalized functions is to be found in Lighthill, M.J., *Fourier Analysis and Generalized Functions*, Cambridge University Press, 1958.

Recall from Equation 2.16 that $\delta(t)$ is defined by its sifting property

$$x(\tau) = \int_{-\infty}^{\infty} x(t)\,\delta(t - \tau)\,\mathrm{d}t.$$

for any value of τ. In order to pick out the value of $x(t)$ at the instant $t = \tau$, the delta function is defined to exist only at $t = \tau$ and to be zero elsewhere. Hence $\delta(t)$ exists only at $t = \tau = 0$ and the defining integral picks out the value $x(0)$ from the signal $x(t)$:

$$x(0) = \int_{-\infty}^{\infty} x(t)\,\delta(t)\,\mathrm{d}t. \qquad (2.17)$$

Now consider the special case where the function $x(t)$ is a constant 1 for all t. Then $x(\tau) = 1$ for any value of τ and we can write the defining integral as

$$1 = \int_{-\infty}^{\infty} \delta(t - \tau)\,\mathrm{d}t = \int_{-\infty}^{\infty} \delta(t)\,\mathrm{d}t \qquad (2.18)$$

Intepreting the integral as a measure of the area under the delta function we reach the conclusion that, although $\delta(t - \tau)$ is defined to exist only at the instant $t = \tau$, the area under the delta function is finite and is equal to unity. No ordinary function has these properties. $\delta(t)$, or $\delta(t - \tau)$ in the general case, is known as the *unit delta function*, or *unit impulse function*.

Remember that an impulse function has a well-defined area, but the height of the function is undefined.

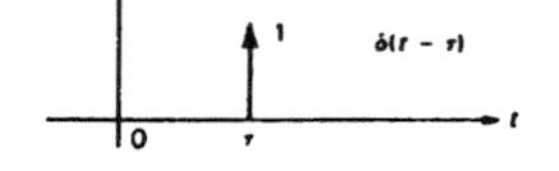

The arrowhead in the conventional representation of the delta function $\delta(t - \tau)$ serves to remind us that the value, or height, of the function at $t = \tau$, is undefined. One way of looking at the delta function is to think of it as the limiting case of a family of pulses, such as the rectangular pulses shown in Figure 2.7. The area enclosed by each pulse is unity. To maintain a constant area the height of the pulses increases as the width decreases. We can imagine the function $\delta(t)$ as being the limiting case as the pulse width becomes smaller and smaller and the pulse height becomes larger and larger.

The unit impulse is a highly idealized function however, so we must now see how we can proceed in practice to determine the impulse response of an actual

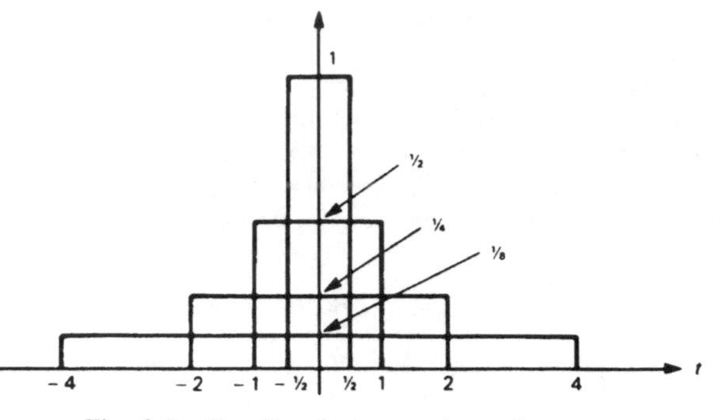

Fig. 2.7 Family of rectangular pulses.

physical system. We shall base our discussion on the simple *RC* network.

One obvious difficulty is that we cannot possibly make any physical quantity, such as a voltage, vary with time exactly as $\delta(t)$ is defined in Equation 2.18. If we try to approximate an ideal impulse input by applying a very large voltage for an extremely short time, we are likely to find that we have exceeded the limits within which our system can be regarded as linear, and the system has either been damaged or has gone into saturation. We overcome this problem by treating the impulse response of a system like the impulse itself – as a limiting case.

The time-constant is a measure of the time taken by a system to respond to an input change. For the *RC* circuit the time-constant is equal to the product *RC*. See Appendix 2 of Ritchie, G.J., *Transistor Circuit Techniques*, Van Nostrand Reinhold, 1987 for a discussion of *RC* network behaviour.

Suppose we apply a voltage pulse of height V and duration T to the input of the *RC* network. We choose the voltage V to be small enough to ensure that the circuit behaves as a linear system. If the pulse width T is greater than the time-constant of the *RC* circuit then the response will be of the form shown in Figure 2.8a. If the pulse width is reduced, while maintaining the same pulse height, the response will change as shown in Figures 2.8b and 2.8c. Because the input pulse area is becoming smaller the size of the response will also decrease. Eventually, however, we will find that the *shape* of the response ceases to change significantly if we continue to reduce the pulse width. Further reduction of the input pulse width will cause a reduction in the size of the response but will have a negligible effect on its overall shape. This occurs when the pulse duration T is very much smaller than the time-constant of the network. We can then model the input as a voltage impulse of area VT and the resulting response of the *RC* network as the decaying exponential

$$v_0(t) = \frac{VT}{RC}e^{-t/RC}. \qquad (2.19)$$

The causal connection between input and response is therefore

$$VT\delta(t) \rightarrow \frac{VT}{RC}e^{-t/RC}. \qquad (2.20)$$

Since $\delta(t)$ is zero everywhere except at time $t = 0$, we imagine that the strength of the impulse is imparted to the system under investigation at the instant $t = 0$.

We define the *unit impulse response* $h(t)$ of the *RC* network by taking the ratio of the response to the strength (or area) of the impulse that caused it.

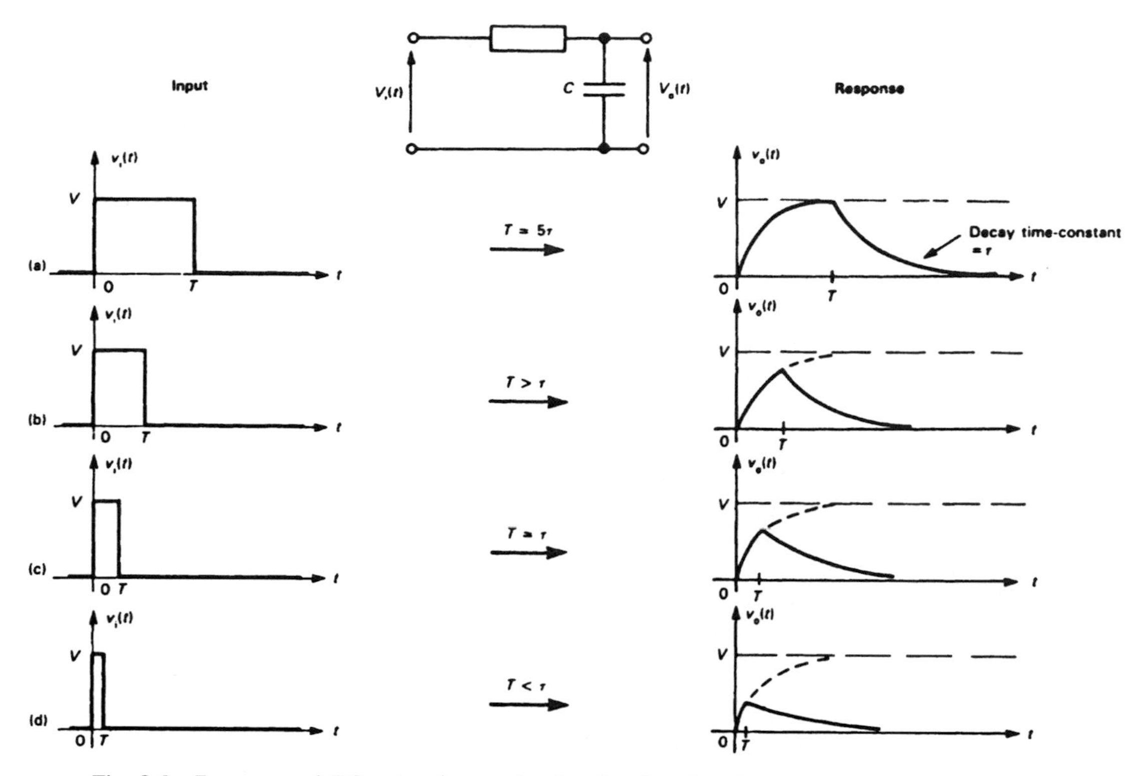

Fig. 2.8 Response of RC network as pulse duration is reduced.

Thus dividing through by VT gives

$$\begin{array}{ccc} \text{input} & & \text{response} \\ \delta(t) & \rightarrow & h(t) = \dfrac{1}{RC} e^{-t/RC} \end{array}$$

as shown Figure 2.9.

The unit impulse response completely describes the dynamic behaviour of a system. Although the impulse itself is a strange concept, the impulse response of a physical system can, in principle, be determined experimentally.

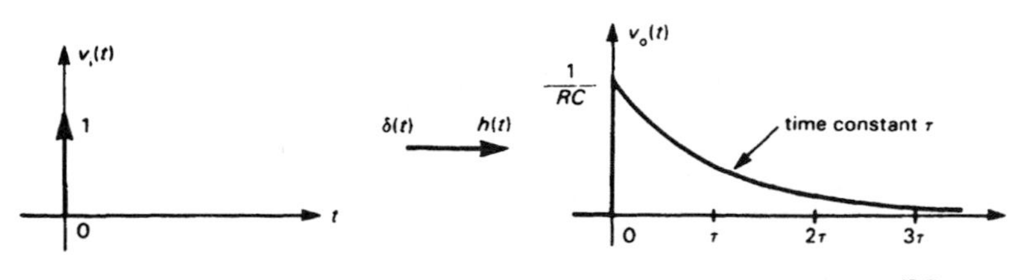

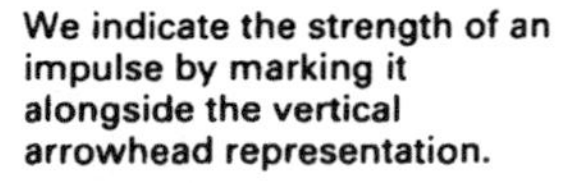

Fig. 2.9 Impulse response of RC network: $\delta(t) \rightarrow (1/RC)\,e^{-t/RC}$

The important feature of an impulse model is its strength, or area, and not its shape. Remember also that the notation $A\delta(t)$ representing a pulse of area A is useful only when the pulse duration is much shorter than the response time of the system under investigation. This means, for example, that a pulse of duration 1 s could successfully be modelled as an impulse for the purposes of calculating the resulting response of a system with a time constant of 100 s. However an impulse model would not be appropriate if we wished to work out the form of the response to the 1 s pulse of a system with a time-constant of 0.1 s.

Worked Example 2.4 The unit impulse response of an RC network has the form

$$h(t) = \frac{1}{RC}e^{-t/RC}, \text{ for } t \geq 0.$$

The time-constant of the network is 1 ms. Use this information to work out the form of the response of the network to a rectangular voltage pulse of amplitude 10 V and duration 10 μs.

Solution: The duration of the input voltage pulse $v_i(t)$ is 100 times smaller than the time-constant of the RC network and hence the pulse can be modelled as an impulse

$$v_i(t) = A\delta(t)$$

where A is the strength of the impulse, defined by the area of the input pulse. This is equal to $10\,\text{V} \times 10 \times 10^{-6}\,\text{s} = 10^{-4}\,\text{Vs}$. Notice that the impulse strength A has the dimensions of [Voltage] × [Time].

The unit impulse response of the network is

$$h(t) = \frac{1}{RC}e^{-t/RC}$$
$$= 10^3 e^{-t/0.001}$$

which has the dimensions of 1/[Time]. The response to an input voltage impulse $10^{-4}\delta(t)$, therefore, is an output voltage of the form

$$10^{-4}h(t) = 10^{-4} \times 10^3 e^{-t/0.001}\,\text{V}$$

which has the dimensions of [Voltage] as required.

The above worked example brings out the important point that when the impulse function $A\,\delta(t)$ is used to model a physical quantity, then we must be aware that both A and $\delta(t)$ have dimensions associated with them. If $A\,\delta(t)$ represents an impulse of voltage, then as we have seen, the impulse strength A, which represents an area under the curve of voltage against time, will have the dimensions [Voltage] × [Time]. For the weighted impulse $A\,\delta(t)$ to have the dimensions [Voltage], therefore, the unit impulse $\delta(t)$ must have the dimensions 1/[Time].

If $A\delta(t)$ represents an impulse of current at time $t = 0$, then A will have the dimensions [Current] × [Time], or charge.

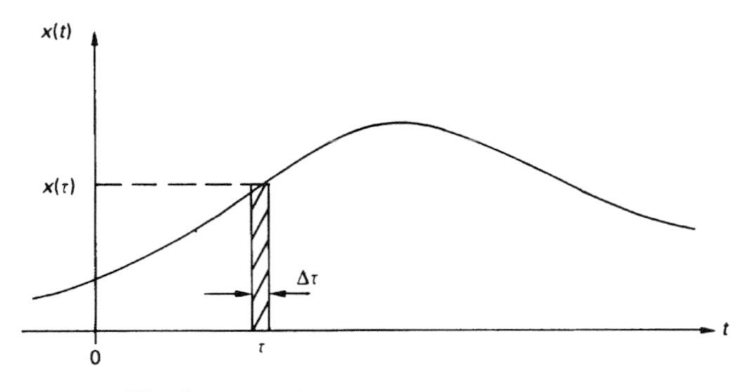

Fig. 2.10 Modelling the input function $x(t)$.

The convolution integral

If we know the unit impulse response $h(t)$ of a continuous system then we can, in principle, work out the response of the system to any input $x(t)$. Consider, for example, the input signal $x(t)$ shown in Figure 2.10. We can approximate $x(t)$ as a succession of rectangular pulses each of width $\Delta\tau$ and height $x(\tau)$, so that the area of a typical pulse centred at $t = \tau$ is $x(\tau)\,\Delta\tau$. If we choose pulse widths that are very small compared to the response time of the system under investigation we can model each pulse as an impulse of the form $A\,\delta(t - \tau)$, where the strength A of the impulse is equal to the area of the pulse. Thus the signal $x(t)$ can be modelled as the continuous sum of impulses, each of the form $[x(\tau)\,\Delta\tau]\,\delta(t - \tau)$. In the limit as the pulse width approaches zero $\Delta\tau$ can be replaced by $d\tau$, and $x(t)$ written as the integral

$$x(t) = \int_{-\infty}^{\infty} x(\tau)\,\delta(t - \tau)\,d\tau \tag{2.22}$$

Since we are dealing with a linear time-invariant system we know that if a unit impulse input $\delta(t)$ gives rise to a response $h(t)$, then a scaled, time-shifted impulse will give rise to a similarly scaled and time-shifted response:

$$[x(\tau)\Delta\tau]\,\delta(t - \tau) \rightarrow [x(\tau)\,\Delta\tau]h(t - \tau). \tag{2.23}$$

Equation 2.22 shows that the input signal $x(t)$ can be expressed as an integral summation of weighted impulses. So, using the principle of superposition we can likewise express the response $y(t)$ of the system to an input $x(t)$ as an integral summation of all the individual impulse responses:

$$y(t) = \int_{-\infty}^{\infty} x(\tau)\,h(t - \tau)\,d\tau. \tag{2.24}$$

This integral is called the *convolution integral* and it defines the continuous convolution of the signals $x(t)$ and $h(t)$. As in the case of its discrete counterpart, the convolution sum, the same output $y(t)$ is obtained if $x(t - \tau)$ and $h(\tau)$ replace $x(\tau)$ and $h(t - \tau)$ in the integral. So we have the equivalent integrals

$$y(t) = \int_{-\infty}^{\infty} x(\tau)h(t-\tau)\,d\tau = \int_{-\infty}^{\infty} h(\tau)x(t-\tau)\,d\tau \qquad (2.25)$$

or, using the * notation to denote convolution,

The convolution integral is sometimes referred to as the 'superposition integral'. In older texts the German term 'Faltung' or 'folding' integral was also used.

$$y(t) = x(t) * h(t) = h(t) * x(t)$$

This means that we will observe the same response from a system irrespective of whether $x(t)$ is the input and $h(t)$ is the impulse response, or $h(t)$ is the input signal and $x(t)$ is the impulse response.

System response calculations

We have established that the input, output and unit-impulse response of a stable linear system are linked in the time-domain by the convolution integral. To round off this chapter the following worked example shows how we can use convolution to work out the response of the *RC* network to a given input.

Worked Example 2.5 Using the convolution integral, work out the response of the *RC* network to an input voltage step of height 1 V.

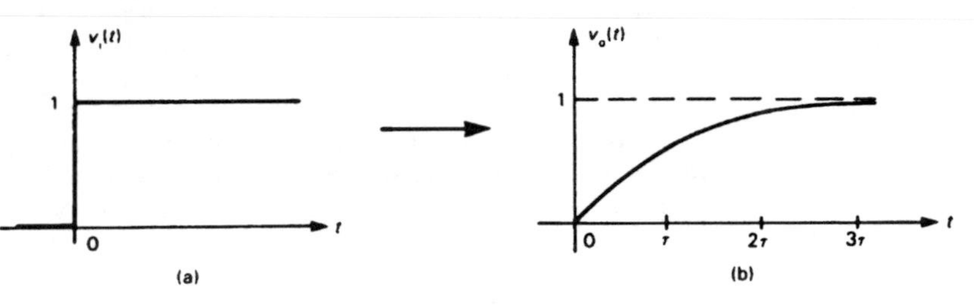

Fig. 2.11 Step response of *RC* network: $u(t) \rightarrow (1 - e^{-t/RC})$.

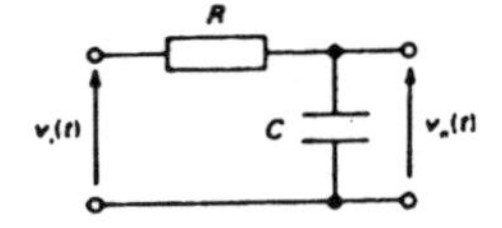

The unit step function $u(t)$ is defined as

$$u(t) = 1 \text{ for } t \geq 0$$
$$= 0 \text{ for } t < 0$$

Solution: We start with the unit step input $u(t)$, shown in Figure 2.11a, and the unit impulse response of the *RC* network, $h(t) = Ae^{-\alpha t}$, where $\alpha = A = 1/RC$.

The response of the system $y(t)$ to the unit step is given by the convolution integral:

$$y(t) = \int_{-\infty}^{\infty} u(\tau)h(t-\tau)\,d\tau.$$

We obtain an expression for $h(t - \tau)$ by substituting $(t - \tau)$ for t in the impulse response, giving

$$h(t-\tau) = Ae^{-\alpha(t-\tau)}.$$

The convolution integral is defined with infinite limits but it turns out that the product $u(\tau)h(t - \tau)$ is zero over some of the range. We can obtain more appropriate limits, therefore, by examining more closely the range over which the integral applies. Firstly, the unit step $u(\tau)$ is defined to be zero for $\tau < 0$. The product $u(\tau)h(t - \tau)$ will therefore be zero for $\tau < 0$, and we can replace the lower limit of integration of $-\infty$ with 0. Secondly, $h(t)$ is the impulse response of a *causal* system and is also defined to be zero for $t < 0$. This means that the function $h(t - \tau)$ will be zero for $t - \tau < 0$, or for $\tau > t$, and that the *upper* limit of integration can be changed from ∞ to t. The convolution integral can thus be written with new limits:

$$y(t) = \int_0^t u(\tau)h(t - \tau)\,d\tau.$$

When we evaluate this integral we obtain an expression for the step response of the RC network as a function of time. Substituting for $u(\tau)$ and $h(t - \tau)$ gives

$$y(t) = \int_0^t 1Ae^{-\alpha(t-\tau)}\,d\tau = \int_0^t Ae^{-\alpha t}e^{\alpha\tau}\,d\tau.$$

Since the term $Ae^{-\alpha t}$ is independent of the variable of integration τ, it can be taken outside the integral to give

$$y(t) = Ae^{-\alpha\tau}\int_0^t e^{\alpha\tau}\,d\tau$$

which can be solved directly to give

$$y(t) = Ae^{-\alpha t}\left|\frac{e^{\alpha\tau}}{\alpha}\right|_0^t = \frac{A}{\alpha}(1 - e^{-\alpha t}).$$

For the RC network, $A = 1/RC$ and $\alpha = 1/RC$, hence the output is

$$y(t) = 1 - e^{-t/RC}$$

which is the expression for the unit step response of the RC circuit. The form of the response is shown in Figure 2.11b.

Using convolution to work out the response of even a simple system like the RC network can become a mathematically cumbersome and tedious process. Because of the difficulties of actually working out the convolution integral, therefore, we seldom use this technique for practical calculations. The convolution relationship, however, provides an important theoretical foundation from which to develop our signal and system models. In the next chapter we shall show how we can use this relationship as a basis from which to move from time-domain to frequency-domain models, and hence to simpler methods of calculation and representation.

Summary

In this chapter we have established the important result that, in the time-domain the response of a linear time-invariant system to an arbitrary input signal is given by the convolution of the input with the impulse response of the system. This result holds for both discrete-time and continuous-time systems. The response of a system to a unit-sample sequence in the discrete case, or to a unit impulse function in the continuous case, characterizes completely the dynamic behaviour of the system.

We introduced two simple discrete-time processors as examples of a non-recursive and a recursive discrete-time system. The unit-sample response of the non-recursive processor contained a finite number of terms, thus defining a finite-impulse response (FIR) system. In contrast, the unit-sample response of the recursive processor contained an infinite number of terms, thus defining an infinite-impulse response (IIR) system.

The unit-sample sequence $\delta[n]$ and the unit-impulse, or Dirac delta, function $\delta(t)$ can be used as basic components in discrete-time and continuous signal models. Both $\delta[n]$ and $\delta(t)$ can be used to 'sift' out the value of a discrete-time or continuous signal at a given value of n or t. The important feature of $\delta(t)$ is that it is defined in terms of its area. In system-response calculations $A\,\delta(t)$ can be used as a model of a pulse where the pulse width is very small compared with the response time of the system under investigation.

Any sequence may be expressed in terms of a sum of scaled and delayed unit-sample sequences. Using the principle of superposition the net response of a system to an arbitrary input sequence is given by summing the responses due to the individual component sequences. This procedure leads to the expression for the convolution sum which relates the response of a discrete-time system to the input and unit-sample response sequences.

Continuous-time signals can be modelled as a continuous sum (an integral) of weighted and time-shifted impulse functions. By analogy with the discrete-time case we express an arbitrary input in terms of weighted and shifted impulse function components. Each input component produces a weighted and shifted impulse response. Summing the individual responses, using an integral instead of discrete summation, gives the net response to the given input signal. This procedure leads to the convolution integral which relates the response of a continuous-time system to the input signal and the unit-impulse response.

Problems

2.1 The unit-step sequence, defined as

$$u[n] = 1 \text{ for } n \geq 0$$
$$= 0 \text{ for } n < 0$$

forms the input to the discrete-time processors described by

(a) $y[n] = \dfrac{x[n] + x[n-1]}{2}$

(b) $y[n] = x[n] + \frac{1}{2}y[n-1]$.

In each case work out the response $y[n]$ to the input sequence.

2.2 For each of the processors in problem 2.1 find the response to a sampled pulse input modelled by the sequence

$$x[n] = 1, 1, 1, 1, 0, 0, 0, \ldots$$

where $x[n] = 0$ for $n < 0$. How do the responses of the processors differ?

2.3 The unit-sample response of a discrete-time system can be written as

$$h[n] = (0.3)^n \text{ for } n \geq 0$$
$$= 0 \text{ for } n < 0.$$

If the input to the system is

$$x[n] = 2\delta[n] - 3\delta[n-2]$$

work out the values of $y[2]$ and $y[5]$.

2.4 The unit-impulse response of a continuous system is

$$h(t) = 3e^{-2t} - 5e^{-4t}.$$

If the input is modelled as

$$x(t) = \delta(t) - 2\delta(t-1) + \delta(t-2.5)$$

find the value of the output $y(t)$ at time $t = 1.5$ s.

2.5 Evaluate the integral

$$\int_{-\infty}^{\infty} f_1(t) \times f_2(t)\,dt$$

where

$$f_1(t) = 2\sin(2000\pi t)$$

and

$$f_2(t) = \delta(t - 0.25 \times 10^{-3}).$$

2.6 An electrical network has the unit-impulse response

$$h(t) = 3t\,e^{-4t}.$$

If a unit voltage step $u(t)$ is applied to the network, use the convolution integral to work out the value of the output after 0.25 seconds.

2.7 A cosine voltage, switched on at $t = 0$, is applied to an RC network with the impulse response

$$h(t) = e^{-\alpha t} \text{ for } t \geqslant 0.$$

The input sinusoid can be expressed as the sum of two exponential components

$$x(t) = \cos\omega_0 t$$
$$\tfrac{1}{2}e^{j\omega_0 t} + \tfrac{1}{2}e^{-j\omega_0 t} \text{ for } t \geqslant 0.$$

Using the convolution integral and the principle of superposition, work out an expression for the response of the RC network. Comment on the form of the response for $t \gg 1/\alpha$.

2.8 Sketch block diagrams for the processors defined by the difference equations:

(i) $y[n] = x[n] + 0.5x[n-1]$
(ii) $y[n] = x[n] - 0.25x[n-2] + 0.5y[n-1]$.

Find the unit sample response for each processor and comment on the result.

2.9 The input sequence $x[n]$ defined in Table 2.1 is applied to a processor defined by

$$y[n] = \frac{x[n] - x[n-1]}{2}.$$

Work out the response $y[n]$ of the processor and explain how it differs from that of the two-term moving averager.

2.10 A steady sinusoidal voltage $v(t) = \cos \omega_0 t$ is sampled at intervals of T seconds to produce the sequence

$$v[n] = \cos \omega_0 nT = \frac{e^{j\omega_0 nT} + e^{-j\omega_0 nT}}{2}.$$

The sequence is applied to a discrete-time processor with the three-term unit-sample response

$$h[n] = 0.5, 1, 0.5, 0, 0, 0, \ldots$$

Using the convolution sum and the principle of superposition, find an expression for the steady-state response of the processor.

Frequency-domain Models 3

Objectives

- ☐ To introduce the idea of representing signals and systems in the frequency domain and to introduce the complex exponential as a basic component in frequency-domain models of signals.
- ☐ To introduce the frequency-response function as the frequency-domain counterpart of the unit-impulse response of a linear system.
- ☐ To introduce the Fourier transform and show how to calculate and interpret the frequency spectrum associated with some simple aperiodic signal models.
- ☐ To describe some of the properties of the Fourier transform and to show in particular the inverse relationship between the duration of a time waveform and the width of its frequency spectrum.
- ☐ To demonstrate that the operation of convolution in the time-domain is replaced by the process of multiplication in the frequency domain.

The time-domain approach discussed in Chapter 2 is only one way of looking at the relationship between signals and systems. In this chapter we are going to present an alternative approach which leads us to think about signals and systems in terms of frequency rather than time. To do this we make use of another rather basic signal component – the complex exponential function $e^{j\omega t}$. Our aim is to develop equivalent frequency-domain representations of some of the more common time-domain signal models. We shall then go on to show how this approach can simplify descriptions of system input–output behaviour by replacing the process of time-domain convolution by multiplication in the frequency domain.

The frequency-domain approach

We shall begin by revising the properties of the *complex exponential* function defined by

$$e^{j\theta} = \cos\theta + j\sin\theta. \tag{3.1}$$

The operator j is defined by the identity $j^2 = -1$. When we are dealing with sinusoidal components, j represents a phase shift of $\pi/2$ radians, or 90 degrees.

If we replace j by $-j$ we obtain the *complex conjugate* expression:

$$e^{-j\theta} = (e^{j\theta})^* = \cos\theta - j\sin\theta. \tag{3.2}$$

If z represents a complex number then the notation z^* is often used to denote the complex conjugate of z.

The complex exponential has a real part, $\cos\theta$ and an imaginary part $\sin\theta$. Using Equations 3.1 and 3.2 we obtain

$$\cos\theta = \tfrac{1}{2}(e^{j\theta} + e^{-j\theta}) \tag{3.3}$$

and

$$\sin\theta = \frac{1}{2j}(e^{j\theta} - e^{-j\theta}). \tag{3.4}$$

The real part of any complex number, or function, z is given by $(z + z^*)/2$. The imaginary part is given by $(z - z^*)/2j$.

Notice that in both cases, the terms appearing on the right-hand side are complex conjugates which combine to represent the real functions $\cos\theta$ and $\sin\theta$.

Worked Example 3.1

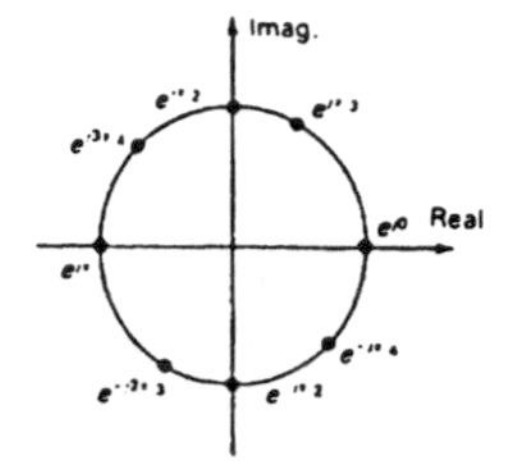

Plot on an Argand diagram the complex numbers represented by $e^{j\theta}$ for the cases where $\theta = 0, \pi/3, \pi/2, 3\pi/4, \pi, -2\pi/3, -\pi/2$ and $-\pi/4$ rads.

Solution: The Argand diagram of the given complex numbers is shown. In each case the *magnitude*, or *amplitude* of the function has a value

$$|e^{j\theta}| = \sqrt{(\cos^2\theta + \sin^2\theta)} = 1.$$

The magnitude of $e^{j\theta}$ is thus equal to 1, irrespective of the value of the phase angle θ, which means that all of its values must lie on a circle of unit radius. Note the special cases: $e^{j0} = +1$, $e^{j\pi/2} = j$, $e^{j\pi} = -1$ and $e^{-j\pi/2} = -j$.

Any complex number may be represented as a point on an Argand diagram. The real part of the complex number defines the horizontal coordinate (abscissa) of the point and the imaginary part defines the vertical coordinate (ordinate).

Signal models

The next step is to put $\theta = \omega t$ in Equations 3.3 and 3.4, where ω represents angular frequency with units of radians per second and t represents time measured in seconds. We then obtain the following models of real signals in terms of complex exponential components:

The terms magnitude, or amplitude, and phase are often used where the complex exponential function occurs in signal analysis. In pure mathematics texts the corresponding terms modulus and argument tend to be more common.

$$\cos\omega t = \tfrac{1}{2}(e^{j\omega t} + e^{-j\omega t}) \tag{3.5}$$

$$\sin\omega t = \frac{1}{2j}(e^{j\omega t} - e^{-j\omega t}). \tag{3.6}$$

If we introduce a scaling factor A we can use Equation 3.5 to express the signal $x(t) = A\cos\omega_0 t$ in the form

$$x(t) = \frac{A}{2}(e^{j\omega_0 t} + e^{-j\omega_0 t})$$

$$= \frac{A}{2}e^{j\omega_0 t} + \frac{A}{2}e^{-j\omega_0 t}. \tag{3.7}$$

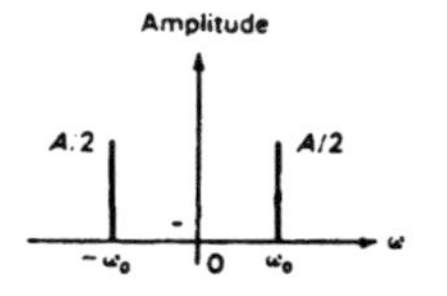

This approach leads directly to the idea of a *frequency-domain* model, where we represent a signal in terms of the amplitude, phase and frequency of its complex exponential components.

Each exponential term is called a *frequency component* of the signal. On the diagram, the amplitudes of the two components of $x(t)$ are represented by lines of height $A/2$ at the frequencies $\omega = \omega_0$ and $\omega = -\omega_0$. Note that one of the components is shown at a positive frequency ω_0 while the other is shown at a negative frequency $-\omega_0$. It is important to remember here that we are dealing with mathematical models of signals and not with the physical signals themselves, where it is not meaningful to distinguish between positive and negative frequencies.

In the general case, a frequency-domain model contains both amplitude and phase information, as illustrated in the following worked example.

Mathematical models like this principally reflect ways in which we find it useful to think about signals; they are not intended to be descriptions of what signals 'really' are.

Worked Example 3.2

A voltage signal is modelled as the sinusoid

$$v(t) = 5\cos(3t + 0.5).$$

Express the signal in terms of exponential frequency components and sketch the frequency-domain representation of the signal.

Solution: The phase-shifted cosine can be written as:

$$v(t) = \tfrac{5}{2}(e^{j(3t+0.5)} + e^{-j(3t+0.5)}).$$

Now the frequency and phase information can be separated by writing the exponential terms in the form

$$e^{j(3t+0.5)} = e^{j3t}e^{j0.5}$$

and

$$e^{-j(3t+0.5)} = e^{-j3t}e^{-j0.5}.$$

Hence we can express the signal as the sum of frequency components at $3\,\text{rad s}^{-1}$ and $-3\,\text{rad s}^{-1}$

$$v(t) = [2.5\,e^{j0.5}]e^{j3t} + [2.5\,e^{-j0.5}]e^{-j3t}.$$

Each component is associated with an amplitude and a phase as illustrated opposite.

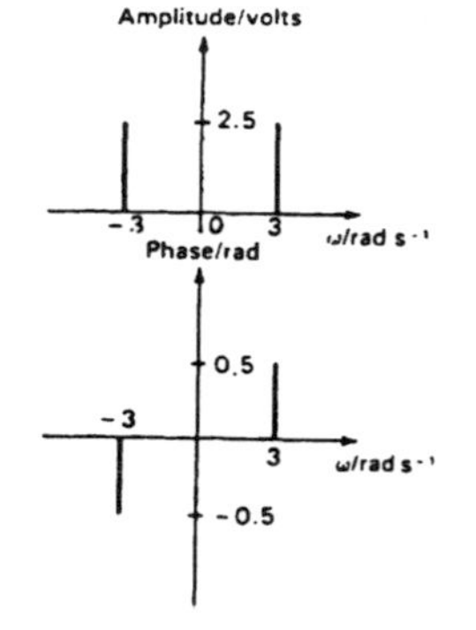

The component at $3\,\text{rad s}^{-1}$ has an amplitude of 2.5 V and a phase of 0.5 rad. The component at $-3\,\text{rad s}^{-1}$ also has an amplitude of 2.5 V but a phase of -0.5 rad.

Exercise 3.1

Work out and sketch the frequency-domain representation of the signal $v(t) = 5\sin(3t + 0.5)$.

System models

The idea that we may represent any signal as a sum of sinusoidal components has important implications for the study of linear systems. As we have seen, if an input signal is represented as the sum of components then the principle of superposition allows us to work out the response of the system to each component separately. We can then add the responses together to find the overall response of the system. We have already met and made use of this result in Chapter 2.

A second point is that the steady-state response of a linear system to a

We can become very adept at looking at signals and systems in frequency-response terms – so much so that we may often feel that this approach is somehow the most 'natural' and that sinusoids really are Nature's own basic building blocks. Remember, however, that sinusoidal components are our own invention and that we find them useful and helpful in modelling the behaviour of certain aspects of the physical world.

Recall that in Chapter 2 we showed that there are two equivalent forms of the convolution integral.

sinusoidal input component is itself a sinusoid of the same frequency as the input but, in general, changed in amplitude and phase. This means that we can develop a model for the behaviour of a system in terms of the effect the system has on the amplitude and phase of sinusoids. It is important to remark that we do this *not* because we are interested in how the system responds to sinusoids in particular, but because such an approach yields a useful description of system behaviour which allows us to relate frequency-domain models of input and output signals in a straightforward manner.

A description of a system in terms of its effect on the amplitude and phase of sinusoidal input components is called a *frequency-response model.*

We know from the previous chapter that the response $y(t)$ of a linear system to an input $x(t)$ is given by the convolution integral

$$y(t) = \int_{-\infty}^{\infty} h(\tau)x(t-\tau)\,\mathrm{d}\tau \tag{3.8}$$

where $h(t)$ is the unit impulse response of the system.

Suppose now that the input $x(t)$ is the general complex exponential component $e^{j\omega t}$. Then $x(t-\tau)$ is equal to $e^{j\omega(t-\tau)}$, or $e^{j\omega t}e^{-j\omega\tau}$, and the convolution integral becomes

$$y(t) = \int_{-\infty}^{\infty} h(\tau)e^{j\omega t}e^{-j\omega\tau}\,\mathrm{d}\tau. \tag{3.9}$$

Because the term $e^{j\omega t}$ is independent of the variable of integration τ it makes no contribution to the value of the integral and hence can be taken outside the integration. The response to the exponential input component is therefore

$$y(t) = e^{j\omega t}\int_{-\infty}^{\infty} h(\tau)e^{-j\omega\tau}\,\mathrm{d}\tau. \tag{3.10}$$

This result gives us all the information we need to determine the effect of the system on a sinusoidal input component at any frequency. Notice that the response $y(t)$ is expressed as the product of two terms: a complex exponential $e^{j\omega t}$ and an integral. We define the integral to be the function of frequency $H(j\omega)$,

$$H(j\omega) = \int_{-\infty}^{\infty} h(\tau)e^{-j\omega\tau}\,\mathrm{d}\tau \tag{3.11}$$

so that the system response is given by

$$y(t) = e^{j\omega t}H(j\omega). \tag{3.12}$$

In other words the response of a linear system to an input component $e^{j\omega t}$ is itself a complex exponential component multiplied by the function $H(j\omega)$.

At any particular frequency $\omega = \omega_0$, $H(j\omega)$ is in general a complex number with a magnitude $|H(j\omega_0)|$ and a phase $\theta(\omega_0)$, where

We have expressed $H(j\omega_0)$ in 'polar' form.

$$H(j\omega_0) = |H(j\omega_0)|e^{j\theta(\omega_0)}. \tag{3.13}$$

For a given input component $Ae^{j\omega_0 t}$, therefore, the system response $y(t)$ will be

$$\begin{aligned} y(t) &= Ae^{j\omega_0 t} \times |H(j\omega_0)|e^{j\theta(\omega_0)}. \\ &= A|H(j\omega_0)|e^{j[\omega_0 t + \theta(\omega_0)]}. \end{aligned} \tag{3.14}$$

The effect of a linear system on a given input component $Ae^{j\omega_0 t}$ is to modify the amplitude of the component by an amount $|H(j\omega_0)|$ and to shift the phase associated with the component by a factor $\theta(\omega_0)$. $H(j\omega)$ is called the *frequency-response function* of the system and is a frequency-domain model of the system.

A linear system does not change the frequency of any input component, or create any new frequency components in the output.

Using the principle of superposition we can quickly work out the effect of a system on an input sinusoid. Specifically, consider an input of the form

$$x(t) = A\cos\omega_0 t$$
$$= \frac{A}{2}e^{j\omega_0 t} + \frac{A}{2}e^{-j\omega_0 t}. \tag{3.15}$$

we can write down the response of the system to each input component separately.

The response to the complex-conjugate component $\frac{A}{2}e^{-j\omega_0 t}$ is the complex-conjugate of the response to $\frac{A}{2}e^{j\omega_0 t}$. In polar form the complex conjugate of $|z|\,e^{j\theta}$ is $|z|\,e^{-j\theta}$.

Hence

$$\frac{A}{2}e^{j\omega_0 t} \rightarrow \frac{A}{2}|H(j\omega_0)|\,e^{j[\omega_0 t + \theta(\omega_0)]}$$

and

$$\frac{A}{2}e^{-j\omega_0 t} \rightarrow \frac{A}{2}|H(j\omega_0)|\,e^{-j[\omega_0 t + \theta(\omega_0)]}$$

So the response to $x(t)$ is the sum of the responses to the individual components

The principle of superposition applies in the frequency domain as well as in the time domain.

$$y(t) = \frac{A}{2}|H(j\omega_0)|[e^{j[\omega_0 t + \theta(\omega_0)]} + e^{-j[\omega_0 t + \theta(\omega_0)]}]$$
$$= A|H(j\omega_0)|\cos[\omega_0 t + \theta(\omega_0)]. \tag{3.16}$$

This shows that the response of a linear system to a sinusoidal input is itself a sinusoid at the same frequency as the input. The amplitude and phase of the response is determined by the input and the value of the frequency-response function of the system at the input frequency.

For a physical system we can find the value of $H(j\omega)$ at any particular frequency $\omega = \omega_0$ by applying a sinewave at that frequency to the input of the system and measuring the amplitude and phase, relative to the input, of the resulting sinusoidal response. For example, if the response to an input $x(t) = A\cos(\omega_0 t + \theta_a)$ is the sinusoid $y(t) = B\cos(\omega_0 t + \theta_b)$, then the magnitude of $H(j\omega)$ at the frequency $\omega = \omega_0$ is given by the *amplitude ratio* of output to input:

We assume that the physical system can be treated as linear for the range of input and output signals in which we are interested.

$$|H(j\omega)|_{\omega=\omega_0} = \frac{\text{output amplitude}}{\text{input amplitude}}$$
$$= \frac{B}{A} \tag{3.17}$$

and the phase of $H(j\omega)$ is

$$\theta(\omega)|_{\omega=\omega_0} = \text{output phase} - \text{input phase}$$
$$= \theta_b - \theta_a. \tag{3.18}$$

These relationships between input and output sinusoids are based on the assumption of a purely sinusoidal input that exists for all times past, present

Transfer-function analysers, which will perform frequency-response measurements automatically over a given frequency range, are available commercially.

and future. To take account of this in practical measurements at a particular frequency we must switch on an input from a sinewave source, a laboratory function generator say, and wait until the system response has settled down to a steady sinusoid of constant amplitude.

Waiting until the output has reached such a *steady-state* ensures that any transient behaviour caused by suddenly switching on the input has effectively been 'forgotten' by the system by the time we come to make our measurement. In general when we model a physical signal as a periodic waveform we do not mean that it has been going on forever – simply that it has been going on long enough to be indistinguishable in practice from an ideally periodic waveform.

Worked Example 3.3

As part of a frequency-response test on an *RC* network, shown in Figure 3.1, a sinewave voltage of the form $10\sin\omega t$ is applied to the input terminals. The steady-state sinusoidal voltage measured across the output terminals has an amplitude of 6 V and lags the input sinusoid by 53 degrees. The test is carried out at a frequency of 100 Hz. What is the (complex) value of the frequency-response function $H(j\omega)$ of the network at this frequency?

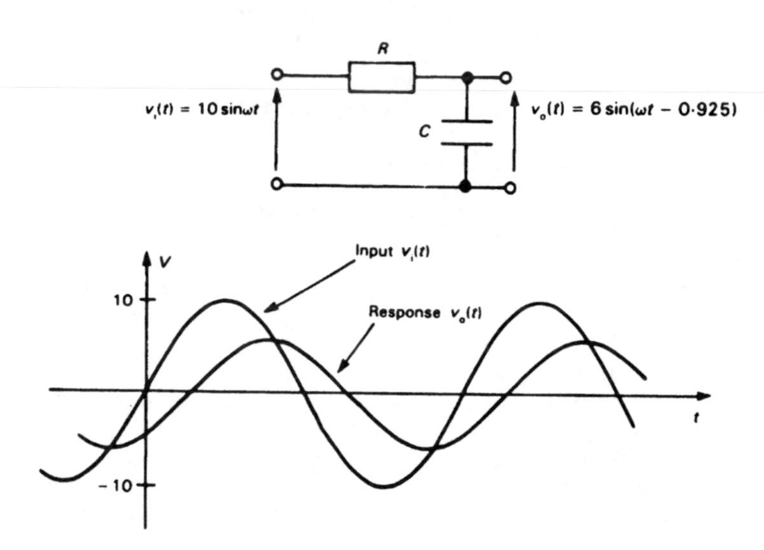

Fig. 3.1 Sinusoidal response of *RC* network.

Solution: The magnitude of $H(j\omega)$ is equal to the amplitude ratio of the output to the input sinusoid. At 100 Hz ($\omega_0 = 200\pi\,\text{rad}\,\text{s}^{-1}$), therefore

$$|H(j\omega_0)| = 6/10 = 0.6$$

The output sinusoid lags the input by 53 degrees, hence the phase of $H(j\omega_0)$ is $-53°$, or -0.925 rad.

At 100 Hz, therefore, the value of $H(j\omega_0)$ is $0.6\,e^{-j0.925}$. Using the identity $e^{-j\theta} = \cos\theta - j\sin\theta$ we can write this as $0.6(\cos 0.925 - j\sin 0.925) = 0.36 - j\,0.48$.

Exercise 3.2

At a frequency of 200 Hz, $H(j\omega)$ has the value $0.12 - j0.33$. If the input is a 10 V sinusoid at this frequency, show that the output will be the voltage $v_0(t) = 3.5\sin(400\pi t - 1.22)$.

Given the form of the frequency response of a linear system, we can find the steady-state output simply by multiplying each input frequency component by the value of the frequency-response function at that frequency. Now, to develop this approach, we must look more closely at the way we model signals and, in particular, at the idea of time- to frequency-domain transformation.

The Fourier transform

The impulse response $h(t)$ and the frequency-response function $H(j\omega)$ are respectively time-domain and frequency-domain models of a system. The important link between them is given by the integral in Equation 3.10.

$$H(j\omega) = \int_{-\infty}^{\infty} h(t)\,e^{-j\omega t}\,dt. \tag{3.19}$$

This integral defines the *Fourier transform* $H(j\omega)$ of the impulse response signal $h(t)$(the change of variable from τ in Equation 3.10 to t does not affect the integral). We can always work out the Fourier integral provided that the function $h(t)$ satisfies the condition

$$\int_{-\infty}^{\infty} |h(t)|\,dt < \infty. \tag{3.20}$$

This is a *sufficient* condition for the Fourier transform of $h(t)$ to exist. When it is satisfied it guarantees that we can work out $H(j\omega)$ from the integral.

In practice this condition is satisfied if $h(t)$ is a pulse-like signal which eventually returns to zero or dies away with time. We say that the integral *converges* under these conditions.

Pulse-like, or finite-energy, signals were discussed in Chapter 1. In general the impulse response of a causal, stable system is a pulse-like signal.

An impulse response which satisfies the condition in Equation 3.20, is the decaying exponential pulse. We met this waveform in Chapter 2 as the impulse response associated with the *RC* network in Figure 3.1. Using the Fourier transform we can immediately work out the corresponding frequency-response function of the network. The exponential pulse is defined as

$$h(t) = e^{-\alpha t} \text{ for } t \geq 0$$
$$= 0 \text{ for } t < 0 \tag{3.21}$$

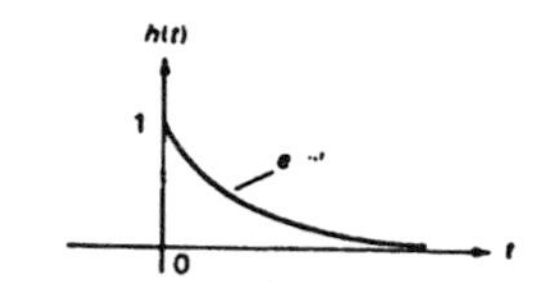

The impulse response of a causal system takes zero values for $t < 0$.

hence the Fourier integral is

$$H(j\omega) = \int_{0}^{\infty} e^{-\alpha t} e^{-j\omega t}\,dt = \int_{0}^{\infty} e^{-(\alpha+j\omega)t}\,dt. \tag{3.22}$$

Since $h(t)$ is zero for $t < 0$ there is no contribution to the integral before $t = 0$ and the lower limit of the Fourier integral is zero. Performing the integration gives

$$H(j\omega) = \left|\frac{-e^{-(\alpha+j\omega)t}}{\alpha + j\omega}\right|_{0}^{\infty} = \frac{1}{\alpha + j\omega}. \tag{3.23}$$

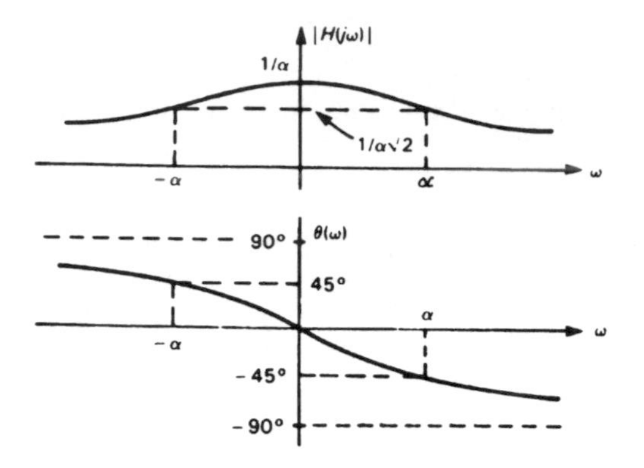

Fig. 3.2 Magnitude and phase of $H(j\omega) = 1/(\alpha + j\omega)$.

This expression is the Fourier transform of the decaying exponential pulse. It is a complex function of frequency ω and contains both magnitude and phase information. The magnitude and phase of the frequency-response function $H(j\omega)$ are given by

$$|H(j\omega)| = \frac{1}{|\alpha + j\omega|} = \frac{1}{\sqrt{(\alpha^2 + \omega^2)}} \quad (3.24)$$

and

$$\theta(\omega) = -\tan^{-1}\frac{\omega}{\alpha}.$$

Figure 3.2 shows the magnitude and phase of $H(j\omega)$ plotted against frequency. Because we can express input components in terms of positive and negative frequencies we consider the frequency range of the network to extend from $\omega = -\infty$ to $\omega = \infty$. There are several points to note about this example. First of all we see that the magnitude response of the network becomes smaller as the frequency increases. This means that the *RC* network will transmit low-frequency input components with relatively little change in amplitude, but will reduce the amplitude of higher-frequency components.

We say that the magnitude response 'rolls off' at high frequencies.

Figure 3.2 also shows how the phase characteristics of the network change with frequency. At low frequencies, the phase shift is relatively small, indicating that the phase of low-frequency input components will be little affected by the network. At higher frequencies, however, the effect of the network becomes correspondingly greater, imposing phase shifts up to a maximum of 90° on very high frequency components.

The overall behaviour of the *RC* network, therefore, is that of a *low-pass filter* where low-frequency signal components are transmitted from input to output with relatively little change, but higher-frequency components are both attenuated and shifted in phase.

In the time-domain the exponentially decaying impulse response of the *RC* network is characterized by its *time-constant* $1/\alpha$. Figure 3.2 shows that there is a simple relationship between the time-constant and the frequency response of the network. At a frequency $\omega = \alpha$ the magnitude of $H(j\omega)$ has fallen to $1/\sqrt{2}$

The time-constant is the time taken by the decaying exponential impulse response to reduce in value by a factor 1/e, or about 0.368.
See, for example, Appendix 2 of Ritchie, G.J., *Transistor Circuit Techniques*, Van Nostrand Reinhold 1987, for a more detailed account of the transient responses of *RC* circuits.

of its value at zero frequency and the system phase shift is equal to $-45°$ (at positive frequency). We use the frequency $\omega = \alpha$ as a point of reference when we refer to the response of the system at high and low frequencies. Low frequencies correspond to $\omega \ll \alpha$ and high frequencies correspond to $\omega \gg \alpha$.

The frequency $\omega = \alpha$ is variously called the 'cut-off frequency', the '3 dB frequency' or the 'half-power point' of the network's frequency response. A sinusoidal input component at this frequency will be reduced in amplitude by a factor of $1/\sqrt{2}$ corresponding to a reduction of power by 1/2, or just over 3 dB $[10\log_{10}(1/2)]$.

The following worked example shows how we can use the above results to determine the frequency-response characteristics of a practical lowpass network.

Worked Example 3.4

The impulse response of an electrical network is modelled by the exponentially-decaying pulse

$$h(t) = 5 \times 10^3 e^{-t/\tau} \quad \text{for } t \geq 0$$
$$= 0 \quad \text{for } t < 0.$$

The time-constant τ of the network, where $\tau = 1/\alpha$, is 1 ms. Work out the response of the network to an input sinewave of amplitude 3 V and frequency 200 Hz.

Scaling by a constant factor A is an equivalent operation in both time and frequency domains. If $H(j\omega)$ is the transform of $h(t)$, then $AH(j\omega)$ is the transform of $Ah(t)$. Fourier transforms, therefore, obey the principle of homogeneity.

Solution: The Fourier transform of the impulse response $Ae^{-\alpha t}$ is $A/(\alpha + j\omega)$ so for $A = 5 \times 10^3$ and $\alpha = 1/\tau = 1000\,\text{rad}\,\text{s}^{-1}$, the frequency-response function of the network is

$$H(j\omega) = \frac{5000}{1000 + j\omega}$$

At a frequency of 200 Hz, ω is equal to $400\pi\,\text{rad}\,\text{s}^{-1}$. Hence

$$H(j\omega) = \frac{5000}{1000 + j400\pi}$$

$$= \frac{5}{1 + j1.26}.$$

from which the magnitude and phase of $H(j\omega)$ are

$$|H(j\omega)| = \frac{5}{\sqrt{(1 + (1.26)^2)}}$$
$$= 3.1$$

and

$$\theta(\omega) = -\tan^{-1}1.26$$
$$= -0.9\,\text{rad, or about } -52°.$$

So the response of the network is a sinusoidal voltage of amplitude $3.1 \times 3 = 9.3$ V lagging the input in phase by 52 degrees. The frequency of the sinusoidal response is the same as the frequency of the input sinusoid.

Fourier transforms of signals

There is no reason why we should restrict ourselves to time-domain signals which are specifically impulse responses of systems. In fact we can work out the

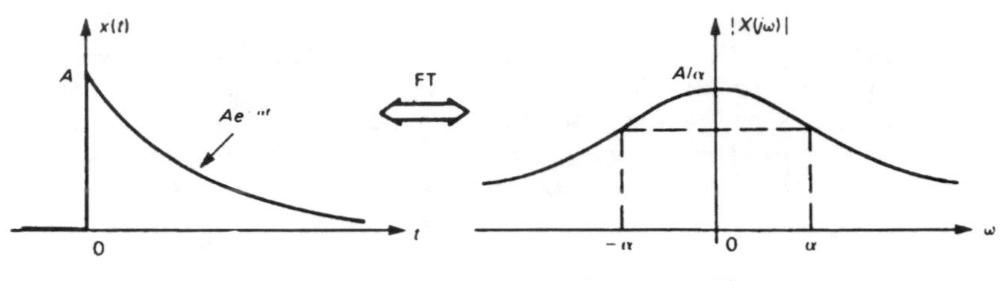

Fig. 3.3 Spectrum of $f(t) = Ae^{-\alpha t}$.

Fourier transform for *any* signal $x(t)$, provided that the conditions for the convergence of the integral are satisfied. In general, then, the integral

$$X(j\omega) = \int_{-\infty}^{\infty} x(t)\,\mathrm{e}^{-j\omega t}\,\mathrm{d}t \tag{3.25}$$

defines the Fourier transform $X(j\omega)$ of the signal $x(t)$.

The upper-case notation $X(j\omega)$ is used conventionally to denote transform of a signal $x(t)$.

The transform $X(j\omega)$, like the frequency-response function $H(j\omega)$, is a continuous function of frequency and at any specific frequency ω its value is a complex number. We interpret the operation of taking the Fourier transform as effectively analysing the signal into a continuous sum of complex exponential frequency components, spread over all positive and negative frequencies. The function $X(j\omega)$ describes how the amplitude and phase of this continuous distribution of components varies with frequency. The Fourier transform $X(j\omega)$ is thus a frequency-domain model of the signal $x(t)$ and is called the *frequency spectrum*, or simply the *spectrum* of the signal.

The relationship between a signal model $x(t)$ and its Fourier transform, or spectrum, $X(j\omega)$ is unique. No two time-domain signals have the same Fourier transform (unless, of course, they are identical) and, similarly, no two transforms correspond to the same time-domain signal. We denote this one-to-one relationship by a double-headed arrow

$$x(t) \leftrightarrow X(j\omega) \tag{3.26}$$

and say that $x(t)$ and $X(j\omega)$ are a *Fourier-transform pair.*

In the previous section we worked out the Fourier transform of an exponentially-decaying pulse of the form $Ae^{-\alpha t}$. We can now interpret the resulting transform as the frequency spectrum of the pulse. In this case we have the Fourier transform pair

$$Ae^{-\alpha t} \leftrightarrow \frac{\mathrm{A}}{\alpha + j\omega}. \tag{3.27}$$

Figure 3.3 shows the isolated pulse and the associated amplitude spectrum $|X(j\omega)|$. This particular example illustrates a number of quite general features of the relationship between the time and frequency domains. Firstly, since $X(j\omega)$ is interpreted in terms of a continuous distribution of components it is meaningless to talk of the contribution of a component at any one particular frequency to the exponential pulse. Instead we should think of $X(j\omega)$ as a *density spectrum.* We say that $X(j\omega)$ gives the *spectral density* of $x(t)$.

If $x(t)$ represents a voltage signal then its Fourier transform $X(j\omega)$ will be

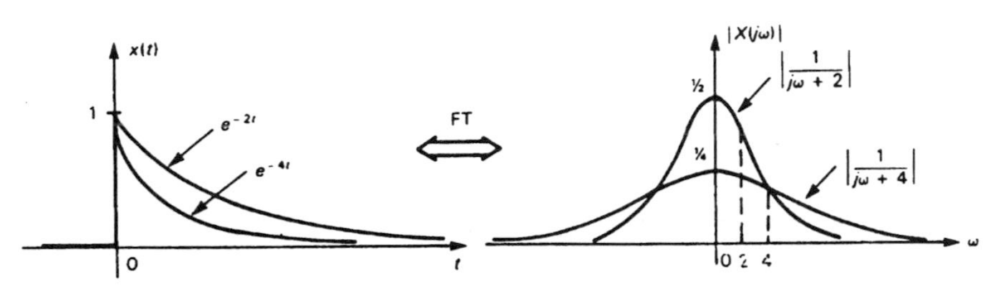

Fig. 3.4 Magnitude spectra of e^{-2t} and e^{-4t}.

a voltage-density spectrum, having the units of volts per unit of frequency. The usual convention is to express the spectral density in units of volts per hertz (VHz^{-1}). If we plot the spectral magnitude of an exponentially-decaying voltage pulse, for example, against frequency in hertz (that is, against $\omega/2\pi$) as shown, then the area under the curve between the two frequencies f_1 and f_2 is expressed in units of volts and is a measure of the contribution of the components in that band of frequencies to the overall signal. Thus the spectrum of the exponential pulse indicates a substantial contribution from relatively low-frequency components around $\omega = 0\,\text{rad}\,\text{s}^{-1}$, whereas the contribution from higher-frequency components over a comparable frequency range is significantly smaller.

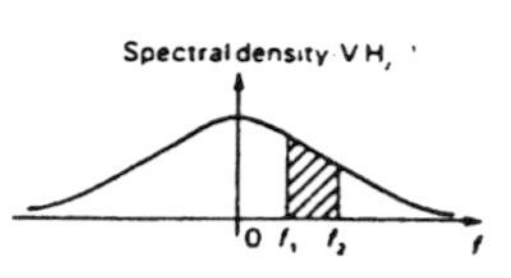

Figure 3.4 shows the amplitude spectra $|X_1(j\omega)|$ and $|X_2(j\omega)|$ of the pulses $x_1(t) = e^{-2t}$ and $x_2(t) = e^{-4t}$, where $x_1(t) = x_2(t) = 0$, for $t < 0$. If we take the time constant $1/\alpha$ of the exponential pulse as a measure of the duration of the pulse in the time domain, and the frequency $\omega = \alpha$ as a measure of the spectral width of the signal in the frequency domain, then we see that the short pulse, $x_1(t)$ is associated with a correspondingly wide spectrum. If the duration of the pulse is increased the spectral width becomes smaller. This is a general principle which applies to the Fourier transform pairs of *all* signals. Thus, any signal of short duration in the time domain will occupy a relatively wide band of frequencies. Conversely a relatively narrow spectrum is associated with a signal of comparatively long duration.

The inverse relationship between time- and frequency-domain behaviour is to be expected since the variable t in $x(t)$ has dimensions of [Time], whereas the frequency variable ω in $X(j\omega)$ has dimensions of 1/[Time].

Figure 3.4 also suggests that for a given pulse shape the overall magnitude of the spectrum is reduced as the pulse duration is reduced. To account for this behaviour, we note that the value of a spectrum at any frequency ω_0 can be found by substituting $\omega = \omega_0$ in the Fourier integral, to give

$$X(j\omega_0) = \int_{-\infty}^{\infty} x(t)\,e^{-j\omega_0 t}\,dt. \tag{3.28}$$

Now, putting $\omega_0 = 0$ and noting that $e^0 = 1$, we obtain

$$X(0) = \int_{-\infty}^{\infty} x(t)\,dt. \tag{3.29}$$

This result shows that the zero frequency value of a spectrum is equal to the net area under the graph of the signal $x(t)$. For the general exponential pulse $Ae^{-\alpha t}$, the area under the curve between $t = 0$ and $t = \infty$ is A/α. Hence for the signals given above, the spectral values are $X_1(0) = 1/2$ and $X_2(0) = 1/4$

This is a useful result to use as a quick check on your working in Fourier transform calculations.

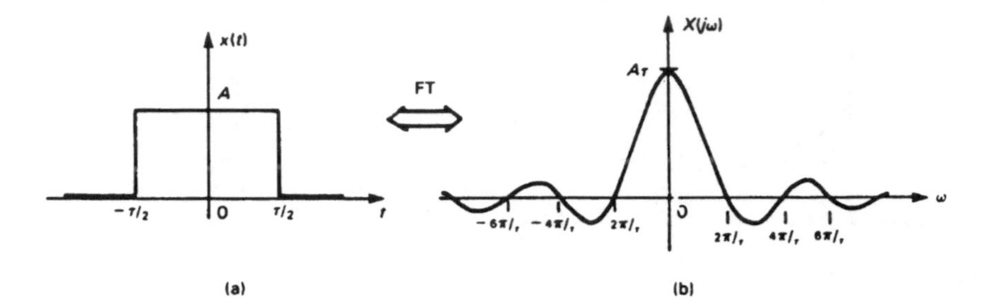

Fig. 3.5 Spectrum of isolated rectangular pulse.

respectively. Clearly we obtain the same values by replacing $\omega = 0$ in their respective Fourier transforms.

We can see more clearly how the relationship between the time- and frequency-domains operates in the case of the rectangular pulse, which we shall discuss next.

Spectrum of the rectangular pulse

More details of the signal models used in communications systems are to be found in O'Reilly, J.J., *Telecommunication Principles*, Van Nostrand Reinhold, 1989.

The rectangular pulse shown in Figure 3.5a is used widely to model the behaviour of signals which switch between 'low' and 'high' levels. In particular such models play an important role in the analysis and design of digital communications systems.

The pulse is of height A, width τ and is symmetrical about the time origin. The model is

$$\begin{aligned} x(t) &= A \quad \text{for } -\tau/2 \leqslant t \leqslant \tau/2 \\ &= 0 \quad \text{elsewhere.} \end{aligned} \tag{3.30}$$

Since there is no contribution to the Fourier integral for $t < -\tau/2$ or $t > \tau/2$, the transform is given by

Recall that $\sin\theta = (e^{j\theta} - e^{-j\theta})/2j$.

$$\begin{aligned} X(j\omega) &= \int_{-\tau/2}^{\tau/2} A e^{-j\omega t}\, dt \\ &= A \left| \frac{-e^{-j\omega t}}{j\omega} \right|_{-\tau/2}^{\tau/2} \\ &= \frac{-A(e^{-j\omega\tau/2} - e^{j\omega\tau/2})}{j\omega} \\ &= \frac{2A\sin\omega\tau/2}{\omega}. \end{aligned}$$

A related function commonly found in texts on telecommunications is the *sinc function*, defined as $\operatorname{sinc} x = (\sin \pi x)/\pi x$.

Multiplying top and bottom of this expression by $\tau/2$ allows us to write this result in the more usual form

$$X(j\omega) = A\tau \frac{\sin \omega\tau/2}{\omega\tau/2}. \tag{3.31}$$

Figure 3.5b shows the spectrum of the single rectangular pulse. We see that $X(j\omega)$ is a wholly real function of ω and this is a direct consequence of the fact that, unlike the decaying-exponential waveform of the previous section, the pulse is symmetrical about the time origin. We shall return to the topic of time and frequency-domain symmetry later.

The spectral density of the pulse is greatest around $\omega = 0$, indicating that the greatest contribution to the signal comes from relatively low-frequency components. At $\omega = 0$ the value of the spectrum is $A\tau$, which is the area under the time-domain pulse. The value of the spectrum is zero when $\sin \omega\tau/2$ is zero. This occurs at the frequencies at which $\omega\tau/2$ is equal to $\pm n\pi$, that is at $\omega = \pm 2\pi/\tau$, $\pm 4\pi/\tau$, $\pm 6\pi/\tau$ and so on. Around these frequencies the spectral contribution to the pulse is least. If we take the frequency of the first zero crossing at $\omega = \pm 2\pi/\tau$ as a measure of spectral width then we find once again the inverse relationship between pulse duration and frequency spread. As the pulse duration τ is made smaller the first zero-crossing point of the spectrum at $\omega = 2\pi/\tau$ moves higher in frequency, and vice versa.

The value of a function of the form $(\sin x)/x$ approaches unity as x approaches zero. Confirm this for yourself by expanding $\sin x$ as a power series in x and then working out the limiting value of $(\sin x)/x$ as x goes to zero.

Exercise 3.3

An isolated rectangular voltage pulse has a height of 5 V and a duration of 1 ms. Work out the value of the spectral magnitude at zero frequency and the (positive) frequency at which the spectrum first becomes zero. What would be the duration of the pulse if the frequency of the first zero crossing were 100 Hz?

The rectangular pulse and its transform are useful building blocks with which to construct Fourier transform pairs of other pulse-like signals. Since the Fourier transform is a linear operation, a signal modelled as the sum of components in the time-domain has a spectrum which is the sum of the spectra of the individual components.

If a signal $x(t)$ is expressed as the sum of simpler components $x_1(t) + x_2(t)$ then the Fourier transform of $x(t)$ is equal to the sum of the transform of $x_1(t)$ and the transform of $x_2(t)$. Try to confirm this result for yourself.

Exercise 3.4

Use the principle of superposition to find the spectrum of the pulse shown opposite. What is the value of the spectrum at zero frequency?

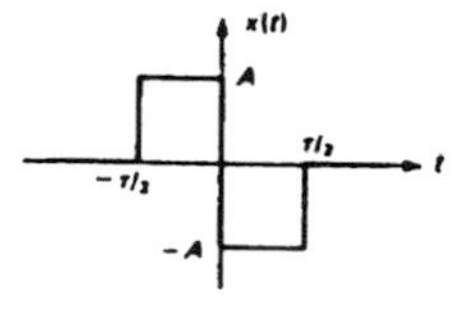

Spectrum of an impulse

Using the idea of the inverse time/frequency relationship we can quickly establish the Fourier transform of the unit impulse or delta function $\delta(t)$ as the limiting case of the transform of a rectangular pulse of height A and width τ.

If the pulse width is reduced but the area $A\tau$ is held constant then the corresponding spectrum will widen, as Figure 3.6 illustrates. Note that the value of $X(j\omega)$ at $\omega = 0$ remains constant and equal to $A\tau$. In the limit, as τ becomes very small, the pulse can be modelled as an impulse of strength $A\tau$ and the associated spectrum becomes a constant of value $A\tau$ at all frequencies. In the case of a unit impulse the strength $A\tau$ is equal to 1, leading to the Fourier transform pair

$$\delta(t) \leftrightarrow 1. \tag{3.32}$$

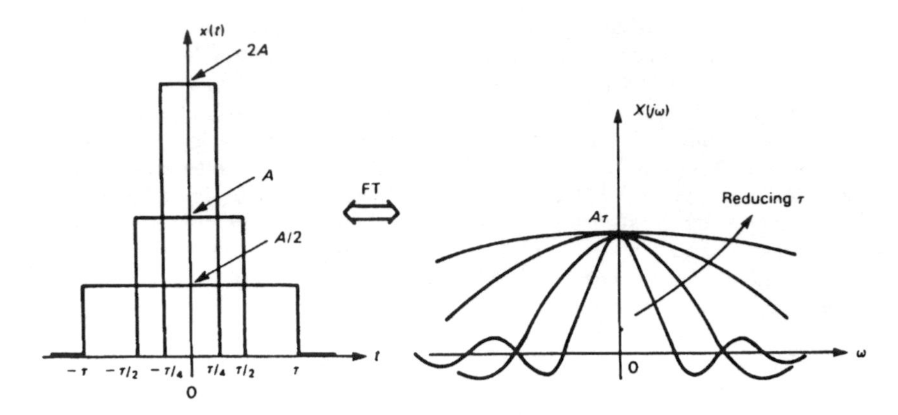

Fig. 3.6 Spectrum widens as pulse width narrows.

It was shown in Chapter 2 that the impulse function, or Dirac delta function, can be thought of as the limit of a sequence of rectangular pulses of constant area but decreasing duration.

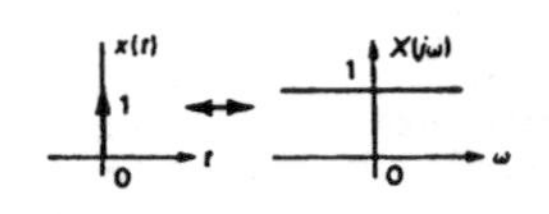

We can use this result to understand the link between the Fourier transform of the impulse response and the frequency-response function of a system. The spectrum of a unit impulse contains components at all frequencies, with equal amplitude and zero phase. When an impulse is used as an input to a system, therefore, we are effectively exciting the system simultaneously at all possible frequencies. The effect of the system is to modify the amplitude and phase of the input spectrum. At each frequency the change in amplitude and phase is given by the value of the frequency-response function of the system. The spectrum of the resulting output, therefore, tells us how the system has modified the uniform spectrum of the input impulse. Hence the spectrum of the impulse response is the actual frequency response of the system.

Spectrum of an oscillatory waveform

In many practical cases we come across pulse-like signals that can be modelled as exponentially-decaying sinusoids. Such a response is often associated with the impulse response of a lightly-damped system. For example, the *RLC* lowpass filter shown in Figure 3.7 will have an oscillatory impulse response of the form shown when $\sqrt{(L/C)} \gg R/2$. The following worked example illustrates some of the features of the spectrum of this type of waveform.

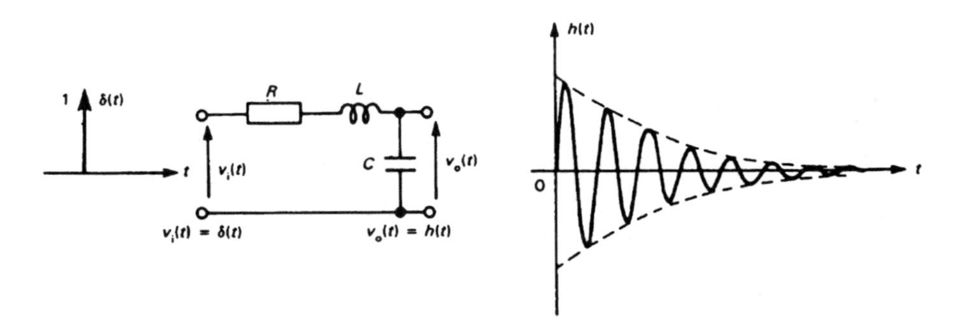

Fig. 3.7 Oscillatory impulse response of lightly-damped *LCR* network.

Worked Example 3.5

An exponentially-decaying sinusoidal voltage has the form $v(t) = Ae^{-\alpha t}\sin\omega_0 t$ for $t \geqslant 0$.

Find the spectrum of this waveform and sketch the form of the spectrum for the case where $\alpha \ll \omega_0$.

Solution: Using the identity $\sin\omega t = (e^{j\omega t} - e^{-j\omega t})/2j$, we can express $v(t)$ in the form

$$v(t) = Ae^{-\alpha t}\frac{(e^{j\omega_0 t} - e^{-j\omega_0 t})}{2j}$$

$$= \frac{A}{2j}e^{-(\alpha - j\omega_0)t} - \frac{A}{2j}e^{-(\alpha + j\omega_0)t}.$$

Using the principle of superposition we calculate the spectrum of each term separately and then combine the results. For the first term the Fourier integral is

$$\frac{A}{2j}\int_0^\infty e^{-(\alpha - j\omega_0)t}e^{-j\omega t}\,dt = \frac{A}{2j}\left|\frac{e^{-(\alpha - j\omega_0 + j\omega)t}}{-(\alpha - j\omega_0 + j\omega)}\right|_0^\infty$$

$$= \frac{A}{2j(\alpha - j\omega_0 + j\omega)}.$$

Similarly the transform of the second term is

Check this result for yourself.

$$\frac{A}{2j(\alpha + j\omega_0 + j\omega)}.$$

Hence, subtracting the transforms gives the overall spectrum of the decaying oscillatory waveform

$$V(j\omega) = \frac{A}{2j(\alpha - j\omega_0 + j\omega)} - \frac{A}{2j(\alpha + j\omega_0 + j\omega)}.$$

$$= \frac{A(\alpha + j\omega_0 + j\omega) - A(\alpha - j\omega_0 + j\omega)}{2j(\alpha - j\omega_0 + j\omega)(\alpha + j\omega_0 + j\omega)}$$

$$= \frac{A\omega_0}{(j\omega)^2 + 2\alpha(j\omega) + (\alpha^2 + \omega_0^2)}$$

$$= \frac{A\omega_0}{(\alpha^2 + \omega_0^2 - \omega^2) + j2\alpha\omega}$$

You may recognize this expression as being very similar to the general form of the frequency-response function of a second-order system.

The resulting spectrum is complex. For the case where $\alpha \ll \omega_0$ the decaying oscillation is lightly damped. In other words the oscillations die away relatively slowly.

The dominant frequency in the response is the frequency of oscillation ω_0. In the spectrum therefore, we would expect to see a strong contribution from frequency components in the region around ω_0. This is confirmed by the plot of the spectral amplitude of the signal, shown in Figure 3.8a. The peaks indicate the predominance of frequency components around $\omega = \pm\omega_0$. The height of the spectral peak will depend on how heavily the oscillation is damped. A very lightly damped oscillation will result in a very high peak. The phase curve in

'relatively slowly' means that over one period $2\pi/\omega_0$ of the oscillation the reduction in amplitude of the decaying sinusoidal response is small.

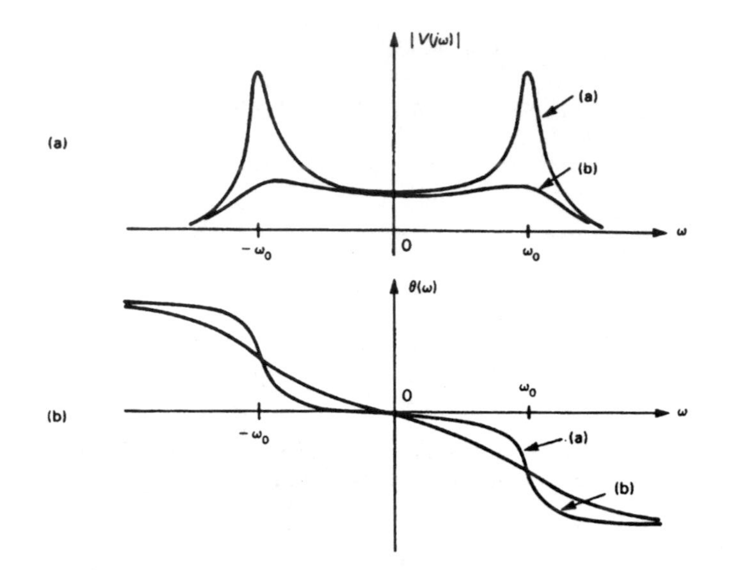

Fig. 3.8 (a) Spectrum of lightly-damped decaying sinusoid; (b) effect of increased damping.

Figure 3.8b shows that, for light damping, the phase of the spectrum changes rapidly around ω_0.

For heavier damping the oscillations of $v(t)$ will die away more rapidly. As a result the height of the spectral magnitude peak will be lower and the phase change around $\omega = \pm\omega_0$ will be more gradual.

Input–output relationships

We established earlier in this chapter that the Fourier transform of the unit impulse response of a linear system gave us the frequency-response function $H(j\omega)$ of the system. That is, $h(t)$ and $H(j\omega)$ are a Fourier transform pair

$$h(t) \leftrightarrow H(j\omega) \tag{3.33}$$

The output of a linear system contains only those frequency components present in the input signal. The system modifies the amplitude and phase of the input components but does not change their frequencies. No components at different frequencies are generated. This is a direct result of the principle of superposition.

Recall that the frequency-response function relates the amplitude and phase of an input frequency component to the amplitude and phase of the corresponding output component. We saw that each spectral component in the output signal is formed by multiplying the corresponding complex exponential input component by the (complex) value of the frequency-response function at the relevant frequency. Remember that because we are dealing with linear systems, each input component is associated with one and only one output component at the same frequency as the input.

Now, we interpret the Fourier transform, or spectrum, of a signal, as a measure of the continuous distribution of the amplitude and phase of the frequency components of the signal. Because of the effect of superposition, all that we have said about the way that linear systems handle single-frequency components can now be applied to the case where an input signal is represented in terms of a continuous distribution of components.

If a signal with a spectrum $X(j\omega)$ is applied to a system the frequency-response function $H(j\omega)$ describes how the system modifies the input spectrum. The spectrum $Y(j\omega)$ of the output signal, therefore, is given by the important relationship

$$Y(j\omega) = X(j\omega)\,H(j\omega) \tag{3.34}$$

This relationship is the frequency-domain equivalent of time-domain convolution. We know that in the time domain the response $y(t)$ to an input $x(t)$ is given by the convolution

$$y(t) = x(t) * h(t)$$

where $h(t)$ is the unit impulse response of the system. Now $y(t)$ and $Y(j\omega)$ are respectively the time- and frequency-domain models of the same signal. Hence they form a Fourier transform pair

$$y(t) \leftrightarrow Y(j\omega)$$

Therefore we can write

$$x(t) * h(t) \leftrightarrow X(j\omega)\,H(j\omega) \tag{3.35}$$

In other words the time-domain convolution of two signals is equivalent to multiplying their spectra in the frequency domain. This is a fundamental result in linear systems theory and indicates the importance of the frequency-domain approach. By working in the frequency-domain we replace the operation of convolution, which is often mathematically very cumbersome, with the simpler operation of multiplication.

Symmetry properties

The Fourier transform of a signal is in general a complex function whose real and imaginary parts depend upon the form of the corresponding time function. In particular they depend on whether the time function is odd or even about the time origin.

A signal model $x(t)$ is an even function of time t if it satisfies the condition $x(t) = x(-t)$. Hence an even function has mirror-image symmetry about the time origin.

We have already seen examples of the relationships that exist. The rectangular pulse in Figure 3.5a, for instance, is an even function of time and its Fourier transform is purely real. On the other hand, a pulse-like signal which is an odd function of time will have a Fourier transform which is purely imaginary. Finally, a function which is neither even nor odd, such as the exponentially-decaying sinusoid in Figure 3.7, has a Fourier transform which has both a real and an imaginary part.

An odd function satisfies the condition $x(t) = -x(-t)$.

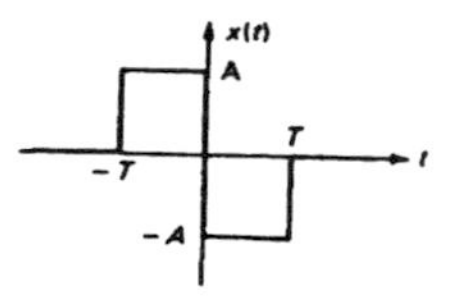

Using the ideas about odd and even functions that were introduced in Chapter 1 we can quickly show that the above results are examples of quite general properties of the Fourier transform. The Fourier integral can be written in terms of its real and imaginary parts

We are using the identity $e^{-j\omega t} = \cos\omega t - j\sin\omega t$.

The notation $R(\omega)$ and $I(\omega)$ indicates that the real and imaginary parts are both *real* functions of ω.

$$X(j\omega) = \int_{-\infty}^{\infty} x(t)\,e^{-j\omega t}\,dt$$

$$= \int_{-\infty}^{\infty} x(t)(\cos\omega t - j\sin\omega t)\,dt$$

$$= \int_{-\infty}^{\infty} x(t)\cos\omega t\,dt - j\int_{-\infty}^{\infty} x(t)\,\sin\omega t\,dt$$

$$= R(\omega) + jI(\omega) \tag{3.36}$$

where $R(\omega)$ and $I(\omega)$ denote respectively the real and imaginary parts of $X(j\omega)$.

Recall that the following relationships hold for the products of odd and even functions:

even × even = even
odd × odd = even
even × odd = odd
odd × even = odd

Suppose now that the time function $x(t)$ is an even function. Then, since $\cos\omega t$ and $\sin\omega t$ are also respectively even and odd functions, the products $x(t)\cos\omega t$ and $x(t)\sin\omega t$ are also respectively even and odd. We know from Chapter 1 that the integral of an odd function over limits symmetrical about the time origin, is zero. Therefore the second integral in Equation 3.36 will be zero and hence the Fourier transform of $x(t)$, where $x(t)$ is even, will be

$$X(j\omega) = \int_{-\infty}^{\infty} x(t)\cos\omega t\,dt = R(\omega) \tag{3.37}$$

which is a purely real function of ω. Sometimes this integral is called the *Fourier cosine integral.*

Similarly, if $x(t)$ is odd, the product $x(t)\cos\omega t$ is odd and the first integral in Equation 3.36 is zero, leading to the *Fourier sine integral*:

$$X(j\omega) = -j\int_{-\infty}^{\infty} x(t)\sin\omega t\,dt = jI(\omega) \tag{3.38}$$

which is a purely imaginary function of ω.

If $x(t)$ is neither even nor odd, as in the case of the exponential pulse or the decaying sinusoid, then both the Fourier sine and cosine integrals have non-zero values and the overall transform $X(j\omega)$ is a complex function of ω. By replacing ω by $-\omega$ in Equation 3.36 we can show that the real part of $X(j\omega)$ is an even function of ω and the imaginary part is an odd function of ω. Hence

$$R(\omega) = R(-\omega) \text{ and } I(\omega) = -I(-\omega). \tag{3.39}$$

Exercise 3.5 Confirm the result that $R(\omega)$ is even and $I(\omega)$ is odd.

The amplitude and phase of $X(j\omega)$ is related to the real and imaginary parts by

$$|X(j\omega)| = \sqrt{[R^2(\omega) + I^2(\omega)]} \tag{3.40}$$

and

$$\theta(\omega) = \tan^{-1} I(\omega)/R(\omega). \tag{3.41}$$

Hence we find that the amplitude of the spectrum of any signal is an even function of frequency, that is

$$|X(j\omega)| = |X(-j\omega)|.$$

The spectral amplitude thus has the same value at $\omega = -\omega_0$ as at $\omega = \omega_0$.

The phase, on the other hand, is an odd function of frequency since

$$\theta(\omega) = -\theta(-\omega). \quad (3.42)$$

Hence the phase at negative frequencies has the same absolute value but is opposite in sign to the phase at positive frequencies.

Using the exponential form of the spectrum we can write these results as

$$X(j\omega) = |X(j\omega)|\,e^{j\theta(\omega)} \quad (3.43)$$

and

$$X(-j\omega) = |X(j\omega)|\,e^{-j\theta(\omega)}.$$

Since $X(-j\omega)$ has the same magnitude but opposite phase to $X(j\omega)$ we recognize that $X(-j\omega)$ is the complex conjugate of $X(j\omega)$. Hence we have the identity

$$X(-j\omega) = X^*(j\omega) \quad (3.44)$$

where * denotes the complex conjugate. In other words the spectrum of a signal at negative frequencies is the complex conjugate of the spectrum at positive frequencies. We say that the Fourier transform of any real signal model has the property of *conjugate symmetry*.

The term 'real' signal model means that the value of $x(t)$ is a real number, as opposed to a complex number, at any time t.

The inverse Fourier transform

The spectrum is a description of a signal in terms of a continuous distribution of complex exponential frequency components. The continuous function $X(j\omega)$ contains all the information about the magnitude and phase of the elementary components $e^{-j\omega t}$ and $e^{j\omega t}$ required to reconstruct the corresponding time-domain signal $x(t)$. Given the spectrum $X(j\omega)$, $x(t)$ can be found from the integral

$$x(t) = \frac{1}{2\pi}\int_{-\infty}^{\infty} X(j\omega)\,e^{j\omega t}\,d\omega \quad (3.45)$$

Recall that if $x(t)$ has the units of volts then $X(j\omega)$ has the units of volts per Hz. Because the variable of integration in the inverse Fourier transform is ω we must divide by the factor 2π to ensure that both sides of the expression for $x(t)$ have the same units of volts.

This integral defines the *inverse Fourier transform*. The factor $1/2\pi$ occurs because we use ω, which has the units of rad s^{-1}, as our frequency variable rather than f, or $\omega/2\pi$, which has the units of Hz.

The inverse Fourier transform can be rather difficult to evaluate analytically, and in fact it is seldom necessary to do so. It is sufficient for our purposes to be aware of the relationships that exist between simple Fourier transform pairs. From a theoretical viewpoint, however, the existence of the inverse transform is important because it completes the symmetry between our time- and frequency-domain models. The existence of the Fourier transform and its inverse shows that the time- and frequency-domain descriptions of a given signal are equivalent and complete representations, where each representation may be obtained from the other.

A more detailed account of the properties of Fourier transforms is to be found in Papoulis. A., *Circuits and Systems – A Modern Approach*, Holt-Saunders, 1980.

Except for a change in the variable of integration (and a scaling factor $1/2\pi$) the inverse transform integral has a similar form to the forward transform. Because of this we might expect that there is some symmetry between the

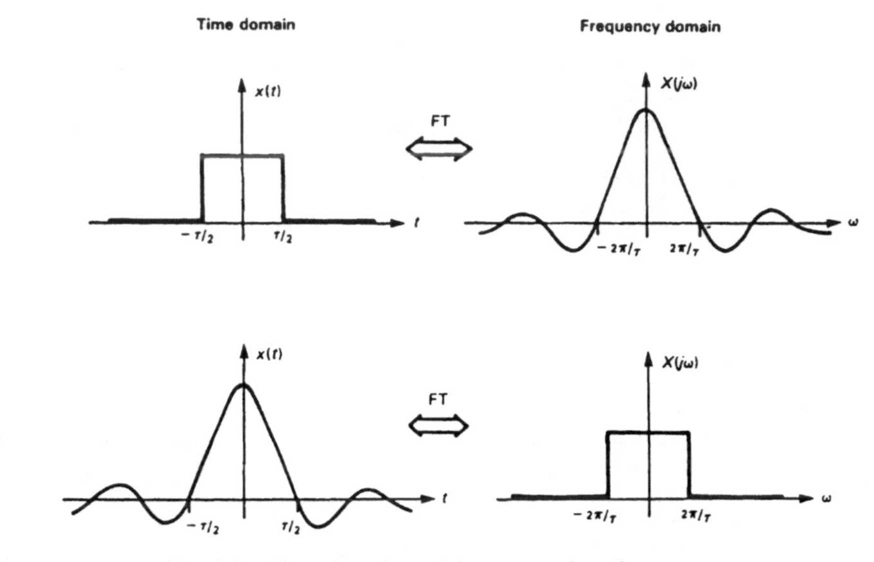

Fig. 3.9 Time-domain and frequency-domain symmetry.

shapes of associated time- and frequency-domain functions. Figure 3.9 illustrates the symmetry between the two domains. Figure 3.9a shows the familiar rectangular pulse and its $\sin x/x$ transform. In Figure 3.9b we see the same effect only now in the reverse direction; a rectangular frequency spectrum is associated with a $\sin x/x$ time-domain pulse. The usual inverse relationship between duration and frequency spread exists in both cases. A narrow $\sin x/x$ pulse is linked with a wide rectangular frequency spectrum and vice versa.

Summary

In the frequency domain we represent a signal as a sum of complex exponential components of the form $e^{j\omega t}$. With each component is associated a magnitude and a phase. In the general case an aperiodic signal is modelled as a continuous distribution of frequency components over all positive and negative frequencies. The variation of the magnitude and phase over the frequency range gives the spectrum of the signal. We calculate the spectrum by working out the Fourier transform of the time-domain signal.

An important property of a linear system is that its response to a complex exponential input component is another complex exponential of the same frequency but modified in magnitude and phase. Since sinusoids can be thought of as being made up of complex exponential components it follows that the steady-state response of a system to an input sinusoid is itself sinusoidal at the same frequency. The ratio of the amplitudes and the difference in phases of the input and output sinusoids is a measure of the frequency-response function of the system at the frequency of interest. In general the frequency-response function is complex function of frequency, and is a frequency-domain model of the dynamic behaviour of the system.

In the time domain the response of a system is given by the convolution of the input signal with the unit impulse response. In the frequency domain the operation of convolution is replaced by the simpler operation of multiplication. The spectrum of the system response is given by the product of the input spectrum and the frequency-response function of the system.

Problems

3.1 A periodic signal voltage is modelled by the expression:

$$v(t) = 3 + \cos 2t + 3\sin(4t + \pi/4).$$

Express this signal in exponential form and hence sketch its frequency-domain representation.

3.2 A first-order linear system is modelled by the frequency-response function

$$H(j\omega) = \frac{3}{1 + j4\omega}$$

(i) Write down an expression for the response of the system to an input signal $x(t) = 5\sin(0.5t + \pi/6)$.

(ii) At what frequency will the magnitude of the frequency-response function have fallen to $1/\sqrt{2}$ of its low-frequency value? What is the phase-shift of the system at this frequency?

3.3 Write down an expression for the unit-impulse response $h(t)$ of the system described in problem 3.2.

3.4 A pulse, modelled as the weighted impulse $2\delta(t)$, is applied to a system and produces the response

$$y(t) = 6e^{-2t} - 4e^{-3t}.$$

What is the frequency-response function of the system?

3.5 The aperiodic signal $x(t)$ and its spectrum $X(j\omega)$ are a Fourier transform pair. Prove the time-shift property of Fourier transforms by showing that a delayed version of the signal $x(t - \tau)$ and its transform form the pair

$$x(t - \tau) \leftrightarrow e^{-j\omega\tau} X(j\omega)$$

3.6 Using the result derived in problem 3.5 find the Fourier transform of the time-shifted impulse $\delta(t - \tau)$. Hence work out the magnitude spectrum of the signal modelled as the impulse sequence

$$x(t) = \delta(t) + 2\delta(t - \tau) + \delta(t - 2\tau).$$

What general feature of the spectrum of $x(t)$ distinguishes it from the spectrum of, say, an exponential pulse?

3.7 The unit-impulse response of an electronic filter can be modelled closely by the expression

$$h(t) = 10^4 e^{-800t} \sin 600t.$$

Use this information to predict the frequency response of the filter and hence sketch the magnitude response on a log frequency scale. At what frequency will the phase-shift of the filter be equal to $-90°$?

3.8 In a communication system, signalling is achieved by keying a sinusoidal signal on and off. One such sinusoidal pulse has a duration T of 20 ms, and is described by

$$
\begin{aligned}
v(t) &= 5\cos 2000\pi t \quad \text{for } -10\,\text{ms} \leqslant t \leqslant 10\,\text{ms} \\
&= 0 \quad \text{elsewhere.}
\end{aligned}
$$

Work out the Fourier transform of this pulse and hence sketch the magnitude spectrum on scaled axes.

3.9 An ideal lowpass filter is modelled as having a rectangular frequency-response function, where $H(j\omega) = 1$ over the frequency range $-\omega_0 \leqslant \omega \leqslant \omega_0$ and zero elsewhere. Use the inverse Fourier transform to work out the unit impulse response of the ideal filter. What does the form of the impulse response indicate about the physical realizability of such a filter?

3.10 The raised-cosine pulse, defined as

$$
\begin{aligned}
x(t) &= 1 + \cos\frac{2\pi t}{\tau} \quad \text{for } \frac{-\tau}{2} \leqslant t \leqslant \frac{\tau}{2} \\
&= 0, \quad \text{elsewhere}
\end{aligned}
$$

is often used for signalling in communications systems. Find the Fourier transform $X(j\omega)$ of the pulse. Hence determine the value of $X(j\omega)$ at $\omega = 0$ and the frequency of the first zero-crossing of the spectrum.

Laplace Transforms 4

Objectives

- ☐ To introduce the Laplace transform and the complex frequency variable s.
- ☐ To show how simple signal models may be represented in terms of s, and to explain the link between the s-domain model and the frequency spectrum of a signal.
- ☐ To show how time-domain operations such as integration, differentiation and delay are represented in the s-domain.
- ☐ To show how the dynamic behaviour of a system may be modelled by a rational function of s.
- ☐ To illustrate how pole-zero diagrams are used to represent graphically Laplace models of signals and systems.
- ☐ To show how pole-zero diagrams are interpreted in both the time domain and the frequency domain.

The Laplace transform builds on the central concept underlying Fourier analysis – that a signal may be modelled equivalently in either the time domain or the frequency domain. The Laplace transform can therefore be thought of as an extension of the Fourier transform, with the advantage that certain signal models, such as the unit step or the continuous sinewave, which are mathematically rather awkward to handle using Fourier methods, may be dealt with more simply.

Non pulse-like signals such as the unit step do not satisfy the conditions for the Fourier transform to converge. We can associate a spectrum with such signals if we allow the Fourier transform to contain frequency-domain delta functions. However this can make the transform difficult to handle in system-response calculations. The Laplace transform avoids these difficulties.

A more compelling reason for introducing another transform, however, is that the Laplace approach leads to a particularly concise pictorial way of representing both the characteristics of signals and the dynamic behaviour of systems. With a little practice we can learn to interpret these models to gain an intuitive insight into the relationships between the time- and the frequency-domain characteristics of signals and systems.

The Laplace integral

The Laplace transform $F(s)$ of the time-domain signal $f(t)$ is defined by the integral

$$F(s) = \int_0^\infty f(t)\,\mathrm{e}^{-st}\,\mathrm{d}t. \tag{4.1}$$

This integral defines the so-called unilateral, or one-sided, Laplace transform.

This integral differs from the Fourier transform in two main ways. First, the integration is from $t = 0$ to $t = \infty$, rather than over all time as in the Fourier case. This means that the Laplace transform $F(s)$ of a signal has built into it the assumption that the signal $f(t)$ is zero before some 'switch-on' instant $t = 0$. As an example, the continuous sinewave $A \sin \omega t$ is treated by the Laplace transform as if it were switched on at $t = 0$.

The assumption that the signal has in effect been switched on at time $t = 0$ is sometimes indicated by writing the signal as $f(t)u(t)$, the product of the signal itself and the unit step function. At any time before $t = 0$ the value of $f(t)u(t)$ is zero. After $t = 0$ the value is simply the value of $f(t)$.

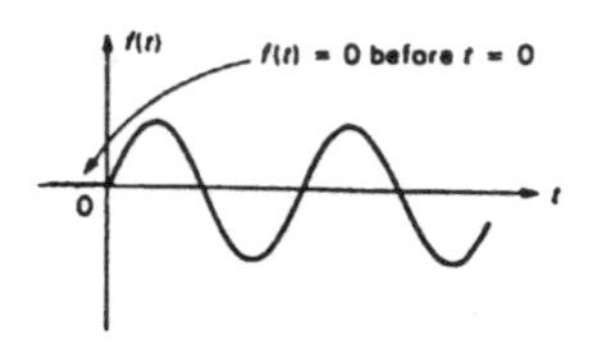

Secondly, the Laplace integral contains the factor e^{-st}, rather than $e^{-j\omega t}$ as in the Fourier case. The Laplace variable s is a complex number, where $s = \sigma + j\omega$. Hence the complex exponential term e^{-st} is equal to $e^{-(\sigma+j\omega)t}$, which can be written as the product $e^{-\sigma t}e^{-j\omega t}$. The term $e^{-j\omega t}$ is identical to the term in the Fourier integral. Depending on the value and sign of σ, the factor $e^{-\sigma t}$ represents a growing or decaying exponential. The product $e^{-\sigma t}e^{-j\omega t}$, therefore, represents an exponentially growing or decaying complex frequency component, where

$$\begin{aligned} e^{-st} &= e^{-\sigma t}e^{-j\omega t} \\ &= e^{-\sigma t}(\cos\omega t - j\sin\omega t). \end{aligned} \tag{4.2}$$

We can therefore think of the Laplace transform as allowing us to model a signal as a continuous distribution of exponentially growing or decaying frequency components of the form e^{-st}, just as the Fourier transform allows us to model signals in terms of a continuous distribution of steady sinusoidal components of the form $e^{-j\omega t}$.

In some textbooks and papers the Laplace variable is represented by p rather than s.

The product st must be dimensionless because raising a number such as e = 2.718... to the power of, say, 2 seconds is meaningless.

The Laplace variable $s = \sigma + j\omega$ is sometimes called the *complex frequency*. This is because s and t appear as a product in the exponential term e^{-st}, and the product st must be a dimensionless complex number for all values of s and t. Since t has the dimensions of [Time], s must have the dimensions of 1/[Time], which are also the dimensions of frequency.

As in the Fourier transform, the imaginary part of s, ω, represents angular frequency and has the units of radians per second. The real part of s, σ, determines the rate of exponential growth or decay of a signal component and has the units of $(\text{seconds})^{-1}$. The following worked example illustrates the part played by $e^{-\sigma t}$ in finding the Laplace transform of the unit step function $u(t)$.

Worked Example 4.1

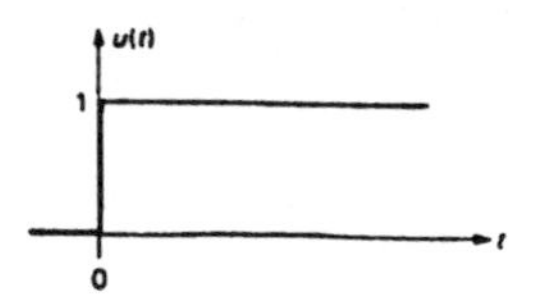

Find the Laplace transform of the unit-step function defined as

$$\begin{aligned} u(t) &= 1 \quad \text{for } t \geq 0 \\ &= 0 \quad \text{for } t < 0. \end{aligned}$$

Solution: To find the Laplace transform we evaluate the integral

$$F(s) = \int_0^\infty u(t)\, e^{-st}\, dt.$$

It is the usual convention to define the lower limit of the integral as the instant just before $t = 0$, and this is sometimes indicated by using the symbol 0_-. We do this in order to handle signal models which have a jump discontinuity at $t = 0$. The value of the signal before time $t = 0_-$ is ignored by the integral. For the unit step we have

$$\begin{aligned} F(s) &= \int_0^\infty 1\cdot e^{-st}\, dt \\ &= \left|\frac{-e^{-st}}{s}\right|_0^\infty \end{aligned}$$

But $s = \sigma + j\omega$, so

$$F(s) = \left| \frac{-e^{-(\sigma + j\omega)t}}{\sigma + j\omega} \right|_0^\infty = \left| \frac{-e^{-\sigma t} e^{-j\omega t}}{\sigma + j\omega} \right|_0^\infty$$

Now at the upper limit, $t = \infty$ the value of the term $e^{-j\omega t}$ is undefined. However the term $e^{-\sigma t}$ (and hence the product $e^{-\sigma t}e^{-j\omega t}$) will be zero for $t = \infty$, provided that $\sigma > 0$. When this condition is satisfied the Laplace transform of $u(t)$ is effectively equal to the integral evaluated at $t = 0$, giving

$$F(s) = \frac{1}{\sigma + j\omega} = \frac{1}{s}$$

for $\sigma > 0$. The function $1/s$ is the Laplace transform of the unit step function $u(t)$.

In the case of the unit step $u(t)$ the Laplace transform converges only when σ, the real part of s, is greater than zero. For a given value of s, therefore, this means that, strictly, the Laplace transform $F(s) = 1/s$ has a defined value only when Re $s > 0$. In practice, however, we treat $F(s)$ as a function of s which has a value for any value of s except, in this example, at $s = 0$ where $1/s$ becomes infinitely large. In the general case we consider that the Laplace transform of a signal exists provided that we can find values of σ, however large, for which the upper limit of the integral goes to zero.

We do not attempt here to present a mathematically rigorous treatment of Laplace transforms or to discuss in detail the conditions under which the integral converges. More formal approaches rely heavily on the theory of complex variables.

It turns out that Laplace transforms exist for virtually all of the deterministic signal models we are likely to come across in signal analysis and processing problems. Unlike the Fourier transform, the Laplace transform can handle not only signals whose amplitudes remain constant after $t = 0$, such as the unit step and the steady sinewave, but also signals that grow with time, such as the linear ramp $f(t) = kt$ (k is the slope of the ramp). We are restricted only in that the growth with time of a particular waveform may not be faster than an exponential growth $e^{\alpha t}$, where α can be any finite number.

A deterministic signal model $f(t)$ is a well-defined function of time. Its behaviour is known for all t.

Exercise 4.1

Work out the Laplace transforms of (a) the exponentially decaying voltage pulse $v(t) = e^{-3t}$; (b) the exponentially growing voltage $v(t) = e^{100t}$.

Laplace models of signals

As with Fourier transforms, the relationship between the time-domain model of a signal and its Laplace transform is unique. Only one Laplace transform can be associated with a given signal and, conversely, there is only one signal corresponding to any particular transform. A signal model and its Laplace transform form a transform pair, which we denote by the double-headed arrow

$$f(t) \leftrightarrow F(s). \tag{4.3}$$

A particularly useful pair is given by the general decaying exponential function $Ae^{-\alpha t}$ and its transform. The Laplace integral is

Uniqueness depends upon the assumption that $f(t) = 0$ for $t < 0$.

As with the Fourier transform the usual convention is to use the upper-case $X(s)$ to denote the transform of the signal $x(t)$. Sometimes the operation of taking the Laplace transform is indicated by L, so that $X(s) = L\{x(t)\}$.

$$F(s) = \int_0^\infty A e^{-\alpha t} e^{-st} \, dt = A \int_0^\infty e^{-(s+\alpha)t} \, dt$$
$$= \left| \frac{-A e^{-(s+\alpha)t}}{s+\alpha} \right|_0^\infty = \frac{A}{s+\alpha}.$$

The resulting transform pair

$$A e^{-\alpha t} \leftrightarrow \frac{A}{s+\alpha} \tag{4.4}$$

provides a starting point from which the transforms of many other functions can be worked out. The transform pair is valid regardless of whether the constants A and α are real or complex numbers. Standard results such as this make it rarely necessary to work out Laplace transforms from first principles using the integral definition. The following worked example shows how we can make use of the transform pair in Equation 4.4 to find the Laplace transform of a sinusoid.

Worked Example 4.2 A sinusoidal voltage is described by $v(t) = 3\cos 4t$. What is the Laplace transform of this waveform?

Solution: Writing the signal in exponential form gives

$$3\cos 4t = \tfrac{3}{2}(e^{j4t} + e^{-j4t})$$

Each exponential term is now replaced by its Laplace transform

$$e^{j4t} \leftrightarrow \frac{1}{s - j4}$$

and

$$e^{-j4t} \leftrightarrow \frac{1}{s + j4}$$

Adding the two transformed terms gives the overall transform of the cosine voltage

$$F(s) = \frac{3}{2}\left[\frac{1}{s - j4} + \frac{1}{s + j4}\right]$$
$$= \frac{3(s + j4 + s - j4)}{2(s - j4)(s + j4)}$$
$$= \frac{3s}{s^2 + 16}$$

The principle of superposition holds for Laplace transforms. If $f_1(t) \rightarrow F_1(s)$ and $f_2(t) \rightarrow F_2(s)$, then $af_1(t) + bf_2(t) \rightarrow aF_1(s) + bF_2(s)$, where a and b are constants.

This example illustrates the point that the Laplace transform is a linear operation. This means that addition and subtraction (and therefore multiplication or division by a constant that is independent of time) are equivalent operations in both the time-domain and the transformed s-domain. As a result, if any signal can be represented as being made up of the sum (or difference) of other simpler waveforms (not necessarily exponentials) then the Laplace transform of the signal is simply the sum (or the difference) of the transforms of the simpler components.

Use this idea in the following exercises (don't work out the integrals!).

The relationship $af(t) \leftrightarrow aF(s)$, where $f(t) \leftrightarrow F(s)$, applies only if a is a constant independent of time. The Laplace transform of $e^{-3t}\cos 4t$, for example, cannot be found by multiplying the transform of e^{-3t} by the transform of $\cos 4t$. Multiplication in the time domain does not lead to multiplication in the s-domain or vice versa.

Exercise 4.2

Use the Laplace transform pairs that we have already established to find the Laplace transforms of these voltage and current waveforms:

(a) $i(t) = 3e^{-2t} + 4e^{-5t}$ (b) $v(t) = 1 - e^{-4t}$

(c) $i(t) = \cos\omega_0 t$ (d) $v(t) = e^{-\alpha t}\sin\omega_0 t$

Remember in (b) that the Laplace transform treats all waveforms as zero before time $t = 0$.

As we have seen in previous chapters, the unit impulse function $\delta(t)$ plays an important role in linear systems theory. The transform of $\delta(t)$ is given directly by the Laplace integral

Remember that the lower limit of the Laplace integral is defined to be 0_-, i.e. just before $t = 0$, so the integral includes the impulse function at $t = 0$.

$$\int_0^\infty \delta(t)e^{-st}\,dt = 1 \tag{4.5}$$

where we have used the sifting property of the impulse function to pick out the value $e^{-st} = 1$ at time $t = 0$. Hence we have the transform pair

$$\delta(t) \leftrightarrow 1. \tag{4.6}$$

Exercise 4.3

Find the Laplace transform of the time-shifted unit impulse $\delta(t - \tau)$.

The inverse Laplace transform

Like the Fourier transform the Laplace transform has an inverse which is formally defined by an integral

$$f(t) = \frac{1}{2\pi j}\int_{c-j\infty}^{c+j\infty} F(s)\,e^{st}\,ds$$

where $c > \sigma$.

However the evaluation of this integral requires a knowledge of contour integration and complex variable theory and is seldom if ever used in routine Laplace transform work.

If $F(s) = L\{f(t)\}$ then $f(t)$ is called the inverse transform of $F(s)$ and is sometimes written as $f(t) = L^{-1}\{F(s)\}$.

So far we have been concerned with finding the transforms corresponding to given time-domain waveforms. We shall now consider the reverse procedure where we want to be able to move from the s-domain back to the time domain.

A frequently-used method is the partial-fraction technique. The method of partial fractions relies on the fact that finding the Laplace transform or its inverse is a linear operation. So if we can split up a transform into a sum of simpler transforms we can find the inverse of the overall transform by finding the inverse of each simpler function separately and then adding them together. Extensive tables of Laplace transform pairs have been prepared by various authors to remove the need to work out a transform or its inverse from first principles. A short table of some of the more commonly used transform pairs is given in the Appendix. We look up the inverse of each partial fraction term in the table and then combine them to find the time waveform corresponding to the original transform. This procedure is illustrated in the following example.

Worked Example 4.3 Find the signal corresponding to the Laplace transform

$$F(s) = \frac{(s+4)}{s(s^2+5s+6)}$$

Solution: The denominator may be written as the product of three factors, s, $(s+2)$ and $(s+3)$, so that

$$F(s) = \frac{(s+4)}{s(s+2)(s+3)}$$

and hence the function can be expressed in partial fraction form as the sum of three terms

$$\frac{(s+4)}{s(s+2)(s+3)} = \frac{A}{s} + \frac{B}{(s+2)} + \frac{C}{(s+3)}$$

where A, B and C are constants. One way to find A, B and C is to multiply both sides of the expression by $s(s+2)(s+3)$, giving

$$(s+4) = A(s+2)(s+3) + Bs(s+3) + Cs(s+2).$$

Substituting $s = 0$ eliminates the terms containing B and C:

$$(0+4) = A(0+2)(0+3)$$

hence

$$A = \tfrac{2}{3}.$$

Similarly, substituting $s = -2$ eliminates the terms containing A and C:

$$(-2+4) = B(-2)(-2+3)$$

hence

$$B = -1.$$

Finally, putting $s = -3$ gives the value of C:

$$(-3+4) = C(-3)(-3+2)$$

hence

$$C = \tfrac{1}{3}.$$

So the partial fraction expansion of $F(s)$ is

Check for yourself that adding the three terms gives the original function.

$$F(s) = \frac{2/3}{s} - \frac{1}{(s+2)} + \frac{1/3}{(s+3)}.$$

Using the table of transform pairs in the Appendix we recognize the term $1/s$ as the transform of the unit step $u(t)$, and the terms $1/(s+2)$ and $1/(s+3)$ as transforms of exponentials. Each term is scaled by the corresponding value of A, B and C, where

$$\frac{2/3}{s} \leftrightarrow \frac{2}{3}u(t)$$

and

$$\frac{-1}{(s+2)} \leftrightarrow -e^{-2t}$$

$$\frac{1/3}{s+3} \leftrightarrow \frac{1}{3}e^{-3t}$$

so that the corresponding time-domain function is

$$f(t) = \tfrac{2}{3}u(t) - e^{-2t} + \tfrac{1}{3}e^{-3t} \text{ for } t \geq 0.$$

Remember that built in to the Laplace transform model of a signal is the assumption that the signal has been switched on at time $t = 0$. Hence the function $f(t)$ is defined to be zero for $t < 0$.

A detailed treatment of the partial-fraction method, including techniques for dealing with repeated factors in the denominator, is to be found in Kuo, F.F., *Network Analysis and Synthesis*, Wiley, 1966, Chapter 6.

Exercise 4.4

Use the partial-fraction technique to find the time-domain waveform corresponding to the transform

$$F(s) = \frac{2(s+2)}{(s+4)(s+1)}.$$

The table of transform pairs given in the Appendix, together with a few simple properties of Laplace transforms that we shall discuss in the next section, are sufficient to deal with a wide range of signals and transforms of practical interest.

A treatment of Laplace transforms and applications, which also includes a large table of transform pairs, is to be found in Spiegel, M.R., 'Laplace Transforms', *Schaum's Outline Series*, McGraw-Hill, 1965. This book also includes many worked examples and exercises.

Properties of Laplace transforms

By exploiting the properties of the Laplace transform we can derive and interpret transforms of signals that are not explicitly listed in tables. We shall briefly discuss three such properties here and show in particular that certain operations such as differentiation and integration with respect to time, become simple algebraic operations in the s-domain.

Time-shift (delay)

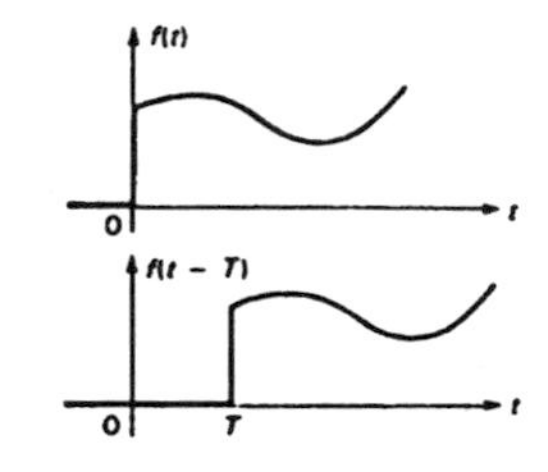

If $f(t - T)$ represents a signal delayed for T seconds, where $f(t - T)$ is assumed to be zero before $t = T$, then the transform of $f(t - T)$ is given by replacing t by $t - T$ in the Laplace integral. Writing $x = t - T$ the integral becomes

$$\int_0^\infty f(x)\,e^{-s(T+x)}\,dx = e^{-sT}\int_0^\infty f(x)\,e^{-sx}\,dx = e^{-sT}F(s). \tag{4.6}$$

Thus if $F(s)$ is the transform of $f(t)$ then the transform of the time-shifted function $f(t - T)$ is $F(s)$ multiplied by e^{-sT}. This property is particularly useful where a signal can be thought of as being made up of the sum of delayed signals, each of which has a known transform.

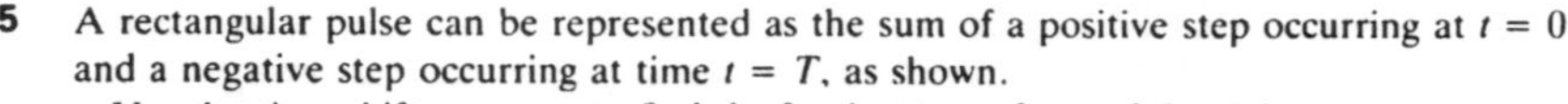

Exercise 4.5 A rectangular pulse can be represented as the sum of a positive step occurring at $t = 0$ and a negative step occurring at time $t = T$, as shown.

Use the time-shift property to find the Laplace transform of the pulse.

Time differentiation

If $f(t) \leftrightarrow F(s)$ then we wish to find the Laplace transform of the time-derivative $\mathrm{d}f(t)/\mathrm{d}t$ in terms of $F(s)$. Using the defining integral and integrating by parts gives

$$\int_0^\infty \frac{\mathrm{d}f(t)}{\mathrm{d}t}\mathrm{e}^{-st}\,\mathrm{d}t = \Big|f(t)\,\mathrm{e}^{-st}\Big|_0^\infty + s\int_0^\infty f(t)\,\mathrm{e}^{-st}\,\mathrm{d}t. \qquad (4.7)$$

In order for the transform $F(s)$ of $f(t)$ to exist, the quantity $f(t)\mathrm{e}^{-st}$ must be zero at $t = \infty$. Hence we have

$$\frac{\mathrm{d}f(t)}{\mathrm{d}t} \leftrightarrow sF(s) - f(0). \qquad (4.8)$$

Thus the transform of the derivative of $f(t)$ is found by multiplying the transform of $f(t)$ by s and subtracting the value of $f(t)$ at time $t = 0$. Recall that for the purposes of transform evaluation we interpret $f(0)$ as $f(0_-)$, that is the value just before $t = 0$. For causal functions which are zero before $t = 0$, such as the unit step, the decaying exponential pulse and the switched-on sinewave, $f(0)$ will be zero. Where this applies, differentiation in the time-domain becomes simply multiplication by s in the s-domain.

By using the time-differentiation property repeatedly, the above result can be extended to higher derivatives of $f(t)$. For example,

$$\frac{\mathrm{d}^2f(t)}{\mathrm{d}t^2} \leftrightarrow s^2F(s) - sf(0) - f'(0) \qquad (4.9)$$

where $f'(0)$ denotes the first derivative of $f(t)$ evaluated at $t = 0$.

The general results for the nth derivative is given in the table in the Appendix. Where $f(0)$, $f'(0)$ and so on are zero this general result shows that repeated differentiation in the time domain turns into repeated multiplication by s in the s-domain.

Integration

Try to confirm the integration result for yourself, noting that the value of the integral of $f(t)$ is zero at $t = 0$.

Integration can be thought of as the inverse of differentiation. It might be expected therefore that this inverse relationship is mirrored in some way in the s-domain. This turns out to be the case. Whereas differentiation becomes multiplication by s, integration becomes division by s. If the transform of $f(t)$ is $F(s)$ then for the integral of $f(t)$ we find

$$\int_0^t f(t)\,\mathrm{d}t \leftrightarrow \frac{F(s)}{s}. \qquad (4.10)$$

Just as for differentiation, repeated integration in the time domain becomes repeated division by s in the s-domain. The Appendix summarizes the Laplace transform properties discussed here. The table also contains a number of other properties that you may find useful for reference.

The system transfer function

Now that we have established some basic properties of the Laplace transform and looked at some simple signal models the next step is to show how the Laplace transform can be used to describe the dynamic behaviour of systems.

Worked Example 4.4

The input and output voltages $v_i(t)$ and $v_o(t)$ of the RC network shown in Figure 4.1a are related by the differential equation

$$CR\frac{dv_o}{dt} + v_o = v_i.$$

If $C = 0.1\ \mu F$ and $R = 100\,k\Omega$ use Laplace transforms to find the response of the network to an input step of height 5 V. Assume the output voltage $v_o(t)$ is zero before the input step is applied, that is $v_o(0) = v_o(0_-) = 0$.

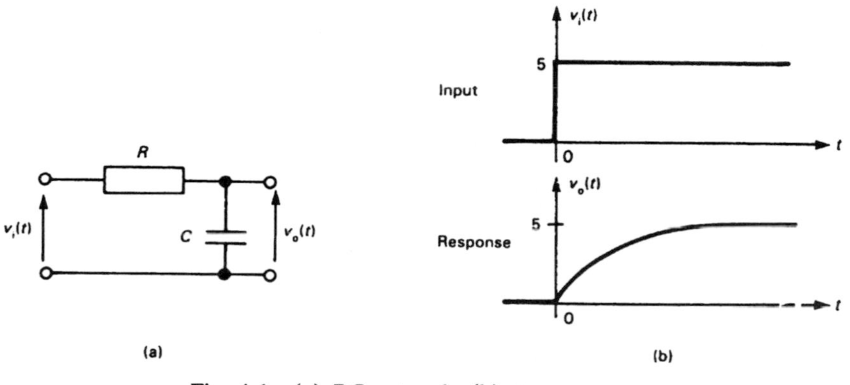

Fig. 4.1 (a) RC network; (b) step response.

Solution: The input and output voltages in the circuit can be represented by their Laplace transforms. Hence we have

$$v_i(t) \leftrightarrow V_i(s) \text{ and } v_o(t) \leftrightarrow V_o(s)$$

and, using the differentiation property where $v_o(0) = 0$,

$$\frac{dv_o(t)}{dt} \leftrightarrow sV_o(s).$$

Replacing each term in the differential equation of the network by its corresponding Laplace transform gives

$$CRsV_o(s) + V_o(s) = V_i(s).$$

The network equation has been transformed into an algebraic expression which we can rearrange to express the transform of the output voltage $V_o(s)$ in terms of the transform of the input voltage $V_i(s)$. Putting

$$(sCR + 1)\,V_o(s) = V_i(s)$$

we obtain the input-output relationship

$$V_o(s) = \frac{1}{sCR + 1} \times V_i(s).$$

The Laplace function $1/(sCR + 1)$ is called the *transfer function* of the network. It links the transform of the output of the network to the transform of the input signal. For a given input voltage waveform, therefore, we can work out the corresponding output voltage by multiplying the transform of the input by the transfer function of the network, and then finding the inverse transform of the result.

In this case the input is a voltage step of height 5 V. That is, $v_i(t) = 5\,u(t)$, hence the corresponding Laplace transform is $V_i(s) = 5/s$.

The transform of the response of the RC network to the input step is therefore,

$$V_o(s) = \frac{1}{sCR + 1} \times \frac{5}{s} = \frac{5}{s(sCR + 1)}.$$

For the given values of C and R, the time-constant of the network, $\tau = CR$, is 0.01 second. Hence the Laplace transform of the network's step response is

$$V_o(s) = \frac{5}{s(0.01s + 1)} = \frac{500}{s(s + 100)}.$$

Compare the procedure for obtaining this result with the procedure necessary to obtain the equivalent result in Chapter 2 using the convolution approach.

This can be expressed in partial fraction form as

$$V_o(s) = \frac{5}{s} - \frac{5}{s + 100}.$$

We recognize the term $5/s$ as the transform of the voltage step $5u(t)$ and the term $5/(s + 100)$ as the transform of the exponential $5e^{-100t}$. Hence we can write down the output voltage as

$$\begin{aligned} v_o(t) &= 5u(t) - 5e^{-100t} \\ &= 5(1 - e^{-100t}) \quad \text{for } t \geq 0. \end{aligned}$$

Figure 4.1b shows the step input and the response of the network.

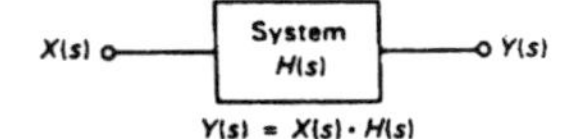

In Laplace calculations it is often assumed that the signal is suddenly applied to a quiescent system at time $t = 0$.

If $X(s)$ and $Y(s)$ represent respectively the Laplace transforms of the input and output signals of a system, then the *transfer function* $H(s)$ of the system is defined as the ratio of the output transform to the input transform.

$$H(s) = \frac{Y(s)}{X(s)}. \tag{4.11}$$

This definition assumes that the system is in a quiescent state, that is the output $y(t)$ and all its derivatives are zero, when the input is applied. The transform of the system response is related to the transform of the input by

$$Y(s) = X(s)H(s). \tag{4.12}$$

The transient part is the response of the system to the sudden change in the input.

When we use Laplace transforms to work out the response $y(t)$ of a system to an input $x(t)$ we find, as in Example 4.4, that, in general, the response can be treated as the sum of two parts: one part has a form determined by the input signal while the other is determined by the *system*. This latter part of the response is often called the *transient response*. In a stable system the transient part of the response will eventually die away. In the Example the transient response of the RC network to the input step was the decaying exponential term $5e^{-100t}$. The time-constant of the decaying transient depends only on the RC time-constant of the network.

The form of the transient response is a property of the *system* and is independent of the input signal.

After a sufficient time the transient part of the response will have become negligibly small and the form of the output of the system will be determined only by the input. When this occurs the system is said to be in a *steady state*. The steady-state response persists for as long as the input persists. In the worked example the steady-state response of the system is the function $5u(t)$.

When frequency-response tests are carried out on a system we must wait for the transient response to die away and the output of the system to settle down to a steady sinewave before we can make any amplitude or phase measurements. Where the transient response is dominated by an exponentially-decaying component of time-constant τ, a good rule of thumb is to assume that the transient part will have become negligibly small after a period of about five time-constants.

When the input is a unit impulse $\delta(t)$ there is a special relationship between the response and the transfer function of a system. The Laplace transform of $\delta(t)$ is 1, so the transform of the impulse response is simply

$$Y(s) = 1 \times H(s) = H(s) \tag{4.13}$$

In other words the transfer function $H(s)$ of a linear system is equal to the Laplace transform of the system's unit impulse response $h(t)$. That is, $h(t)$ and $H(s)$ are a Laplace transform pair:

$$h(t) \leftrightarrow H(s). \tag{4.14}$$

Recall that we found a similar relationship between the Fourier transform of $h(t)$ and the frequency-response function $H(j\omega)$ of a linear system.

Pole-zero models

The transfer function $H(s)$ of a linear time-invariant system is, in general, a rational function of the Laplace variable s; that is, it is a ratio of polynomials in s. Such a function may be completely characterized, within a constant multiplier, by the values of s at which $H(s)$ becomes infinitely large or zero. These values of s locate the *poles* and *zeros* of $H(s)$. The poles of a rational function occur at those values of s at which the denominator becomes zero, and the zeros occur at those values of s which make the numerator (and hence the entire function) equal to zero. As an example consider the transfer function

$$H(s) = \frac{s + 4}{(s + 2)(s + 3)} \tag{4.15}$$

In this example $H(s)$ also approaches zero as s becomes very large. In the limit as s approaches ∞ the value of $H(s)$ is zero. We describe this by saying that $H(s)$ has a 'zero at infinity'.

In this case the value of $H(s)$ is zero where $s = -4$, and is infinitely large when the denominator is zero, that is where $s = -2$ and $s = -3$. We say that $H(s)$ has a zero at $s = -4$ and poles at $s = -2$ and $s = -3$.

Conventionally, zeros at infinity are not indicated on the pole-zero diagram.

We represent the poles and zeros of $H(s)$ by marking the values of s at which they occur on an Argand diagram, or *pole-zero diagram*. Recall that s is a complex variable, where $s = \sigma + j\omega$. We plot the real part of s, σ, along the horizontal axis and the imaginary part, ω, along the vertical axis. Any particular value of s, therefore, is represented by a point on the diagram. The values of s at which the poles of a function exist are indicated by crosses, as at $s = -2$

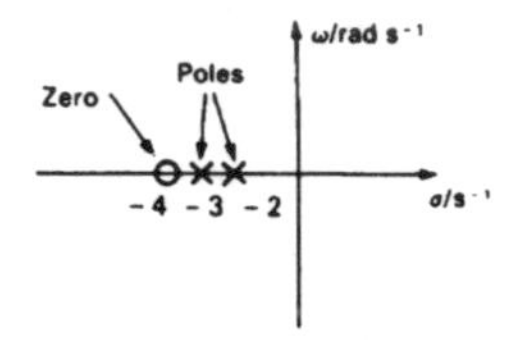

The plane of all possible values of the complex variable s is called the 's-plane'.

The poles and zeros of a physical system can occur only at real values of s, or in complex-conjugate pairs. This is a useful check when working with transforms. If isolated poles or zeros turn up at complex values of s then there is almost certainly an error somewhere in the calculation.

and $s = -3$ in this example. Similarly the zeros of a function are indicated by small circles, as at $s = -4$.

The general form of a transfer function is

$$H(s) = \frac{b_m s^m + b_{m-1}s^{m-1} + \ldots + b_1 s + b_0}{a_n s^n + a_{n-1}s^{n-1} + \ldots + a_1 s + a_0} \tag{4.16}$$

where the coefficients $b_0, b_1, \ldots b_m$ and $a_0, a_1, \ldots a_n$ of the numerator and denominator polynomials are real numbers. Alternatively, $H(s)$ can be written in factored form

$$H(s) = \frac{K(s - z_1)(s - z_2)\ldots(s - z_m)}{(s - p_1)(s - p_2)\ldots(s - p_n)} \tag{4.17}$$

where K is a constant multiplier, and the z_i and p_i are respectively the zeros and poles of $H(s)$. Note that knowledge only of the poles and zeros of a function does not specify the function uniquely. The value of K, which is not represented on a pole-zero diagram, is needed for a complete description.

Since the coefficients of the numerator and denominator polynomials are real numbers, the poles and zeros must either be real, as in the example we have been discussing, or occur in complex-conjugate pairs.

Interpretation of the pole-zero diagram

We have seen that the characteristics of both signals and systems may be represented by functions of s. Hence both signals and systems may be represented by pole-zero diagrams. With a little practice we can learn to interpret a given pattern of poles and zeros in terms of the time-domain and the frequency-domain behaviour of the corresponding signal or system.

Time-domain interpretation

We have seen that any rational function of s can be expressed as a sum of partial-fraction terms. The terms are related to the poles of the function. So, for example, the function

$$F(s) = \frac{3s^2 + 18s + 14}{(s + 2)(s^2 + 2s + 10)} \tag{4.18}$$

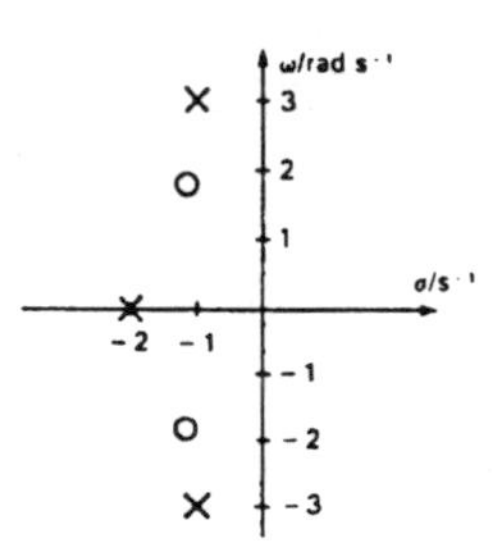

which has the pole-zero diagram shown, can be expressed as the sum of two terms

$$F(s) = \frac{1}{(s + 2)} + \frac{2(s + 1)}{(s^2 + 2s + 10)}. \tag{4.19}$$

Check for yourself that the partial fraction expansion is equivalent to the original function.

The first term is associated with the pole on the real axis at $s = -2$, and the second term with the complex-conjugate pair of poles at $s = -1 + j3$ and $s = -1 - j3$. The time-domain function associated with $F(s)$ is the sum of the functions associated with each of the partial-fraction terms. In this case we have

$$f(t) = e^{-2t} + 2e^{-t}\cos 3t. \tag{4.20}$$

The real-axis pole, therefore, corresponds to a decaying exponential component, while the complex-conjugate pair of poles corresponds to an exponentially-

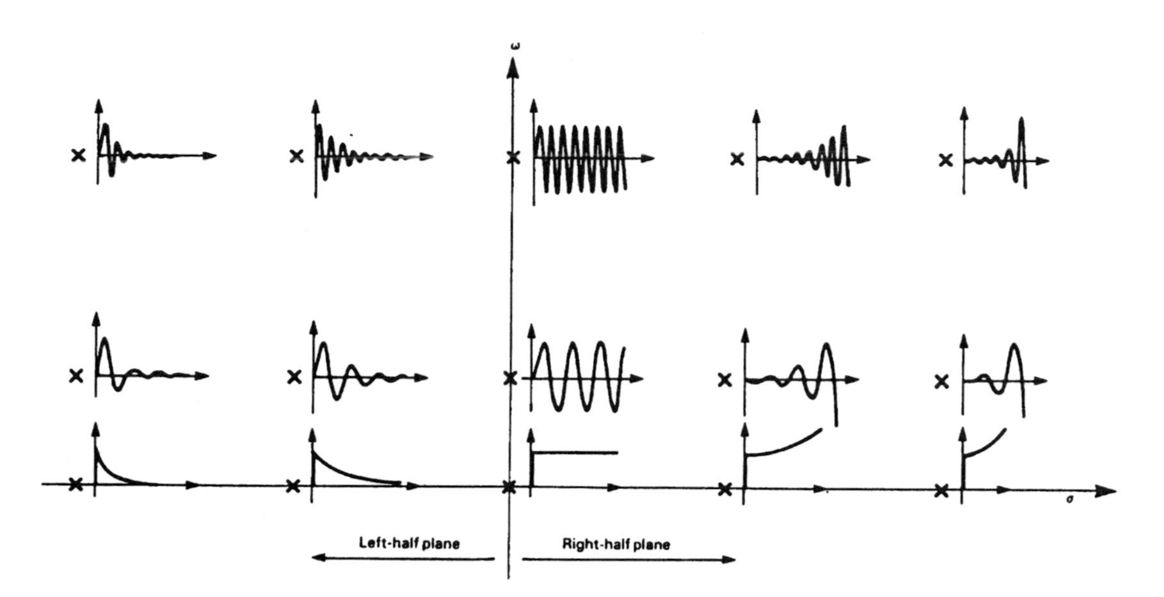

Fig. 4.2 Time waveforms corresponding to s-plane pole locations (only the upper-half of the s-plane is shown).

decaying sinusoidal component. Note that it is the poles that determine the form of the time-domain components. The zeros, on the other hand, do not appear explicitly in the partial-fraction expansion, but determine the relative contribution of each of the components to the overall signal $f(t)$.

If a pole-zero diagram represents the Laplace transform of a signal then we interpret each pole, or pair of poles, as representing an exponential component in the time-domain model of the signal. If the diagram represents the transfer function of a system then the poles represent the time-domain components in the system's impulse response. Figure 4.2 summarizes the relationship between pole positions and the form of the corresponding time-domain component. The diagram shows only half of the s-plane – the part above the real axis. Each pole not on the real axis represents one of a complex conjugate pair. The waveform shown next to each pole is the one associated with that pole, or with the complex conjugate pair.

The impulse response is a special case of the general transient behaviour of a system. In general the transient response of a system is the weighted sum of the time components represented by the poles. A pole at $s = \sigma_0 + j\omega_0$ corresponds to a complex exponential term $e^{-(\sigma_0 + j\omega_0)t}$ in the time domain.

The waveforms associated with poles that lie in the region to the left of the imaginary axis of the s-plane all die away with time. This region is called the *left-half plane*. Waveforms that are associated with poles that lie to the right of the imaginary axis, in the *right-half plane*, grow exponentially with time. Hence the impulse response of any system that possesses right-half plane poles will grow indefinitely, indicating an *unstable* system. For a system to be stable *all* of its poles must lie in the left-half plane.

The topic of stability was introduced in Chapter 1. Only one system pole need lie in the right-half plane for the entire system to be unstable, because the growing exponential associated with the pole will dominate the response of the system.

Poles that lie in the left-half plane on the negative real (σ) axis correspond to non-oscillatory exponentially-decaying components. A pole at $s = -\sigma_0$, therefore, corresponds to a component of the form $e^{-\sigma_0 t}$. The further to the left a pole lies, the larger will be the value of σ_0 and hence the faster the corresponding exponential component will decay. Poles that lie on the positive real axis, in the right-half plane, correspond to non-oscillatory exponentially-growing components of the form $e^{\sigma_0 t}$. The further to the right the pole lies the faster the exponential time-function will grow.

A pole at the origin $s = 0$ corresponds to the unit step-function $u(t)$. The pole lies on the boundary between left- and right-half planes and, hence, the step function represents the boundary between an exponentially-growing and an exponentially-decaying component.

Poles that lie in complex-conjugate pairs in the s-plane are associated with exponentially growing or decaying sinusoidal components. A pair of left-half plane poles at $s = -\sigma_0 + j\omega_0$ and $s = -\sigma_0 - j\omega_0$ correspond to an exponentially-decaying component of the form $e^{-\sigma_0 t} \sin(\omega_0 t + \theta)$(where θ is a constant). The real part of the pole position, σ_0 determines the rate at which the sinusoidal component decays, and the imaginary part, ω_0 determines the frequency of the oscillation. Moving the poles to the left corresponds to making σ_0 more negative, and hence the oscillatory component will die away faster. Poles close to the ω, or imaginary, axis correspond to lightly damped, slowly decaying sinusoidal components. Moving the poles vertically corresponds to changing the value of ω_0. Poles close to the real axis thus correspond to relatively low-frequency, exponentially-decaying, sinusoidal components, while poles at higher values of ω_0 correspond to higher-frequency components.

Complex-conjugate pairs of poles in the right-half plane correspond to exponentially-growing sinusoids of the form $e^{\sigma_0 t} \sin(\omega_0 t + \theta)$. Such waveforms are usually associated with the time-domain behaviour of unstable systems.

The imaginary axis is the dividing line between growing and decaying components. Poles which lie on the imaginary axis are associated with signal components that maintain a constant amplitude for all time after $t = 0$. A pair of poles at $s = +j\omega_0$ and $s = -j\omega_0$ corresponds to a steady sinewave of the form $\sin(\omega_0 t + \theta)$. The frequency of the sinewave is determined by the positions of the complex-conjugate pair of poles on the imaginary axis. Hence a low-frequency sinusoidal component is indicated by poles close to the origin of the s-plane, $s = 0$, while higher frequencies are associated with imaginary-axis poles that lie further away.

Frequency-domain interpretation

A transfer function or signal transform has a well-defined value at any particular value of s, except at the poles. By replacing s by $j\omega$ we can work out the value of a given function of s at points along the imaginary axis, or ω-axis, in the s-plane. As we saw above, points along this axis correspond to steady sinusoids. Putting $s = j\omega$ in a Laplace signal model means that we are expressing the signal in terms of its steady sinewave components. In the case of the isolated exponential pulse $e^{-\alpha t}$, for example, replacing s by $j\omega$ in the transform $1/(s + \alpha)$ gives

Remember that the Laplace transform treats all signals as if they were switched on at $t = 0$. Hence putting $s = j\omega$ in the transform (assuming that it is valid to do so) gives the spectrum of a switched-on signal. Rapid changes in a signal, such as those caused by switching, give rise to high-frequency components in the spectrum. So we would expect to find that the spectrum found from the Laplace transform of a signal extends to relatively high frequencies.

$$F(j\omega) = \frac{1}{j\omega + \alpha} \tag{4.21}$$

which we immediately recognize as the *Fourier transform*, or *spectrum*, of the pulse.

Replacing s by $j\omega$ in the Laplace transform of a signal $f(t)$ to give the spectrum $F(j\omega)$ is a straightforward operation for many of the signal models we come across, but we must be careful not to apply the technique blindly. In particular, we can talk about the spectrum of a signal in a physically meaningful way only where the Fourier transform of the signal exists. If we cannot work out the Fourier integral of a given signal then replacing s by $j\omega$ in the Laplace transform gives an expression which cannot be interpreted physically.

If we replace s by $j\omega$ in the transfer function of a system $H(s)$ we obtain the system's frequency-response function $H(j\omega)$. In this case this procedure is valid only if the system is stable. That is, $H(s)$ has no poles in the right-half of the s-plane. The frequency-response function $H(j\omega)$ gives the steady-state response of the system to an input sinewave.

If the system is unstable then the output will grow indefinitely and will never settle to a steady sinusoidal response. In such circumstances it is impossible to make frequency-response measurements directly. For systems with right-half plane poles, therefore, the associated function $H(j\omega)$ cannot be interpreted in terms of directly measureable behaviour.

The Laplace approach can handle signals for which the Fourier integral does not converge. So, for example, the exponentially growing signal $f(t) = e^{2t}$ has a Laplace transform $F(s) = 1/(s-2)$, but the Fourier transform of $f(t)$ does not exist. Hence although we can replace s by $j\omega$, we cannot interpret the resulting expression $F(j\omega) = 1/(j\omega - 2)$ as a physical frequency spectrum.

Exercise 4.6

A system is modelled as a pure time-delay such that an input signal $x(t)$ produces a response $x(t-T)$. Find the transfer function $H(s)$ and hence the frequency-response function $H(j\omega)$ of the system.

We can learn to interpret a pole-zero diagram to gain an intuitive 'feel' for the form of a spectrum or frequency-response function. We do this by visualizing how the poles and zeros influence the value of $H(s)$ at different values of s. At any particular value of s (except at the poles), $H(s)$ is a complex number which we can express in terms of a magnitude $|H(s)|$ and a phase $\theta(s)$. We will concentrate only on the magnitude characteristics of $H(s)$.

Consider as an example the function $H(s) = s/(s^2 + s + 1)$, which has a zero at $s = 0$ and a complex-conjugate pair of poles at $s = -1/2 + j\sqrt{3}/2$ and $s = -1/2 - j\sqrt{3}/2$, as shown in Fig. 4.3a. The magnitude function is

$$|H(s)| = \left|\frac{s}{s^2 + s + 1}\right| \tag{4.22}$$

We can represent the magnitude of $|H(s)|$ as the height of a surface above the s-plane, as shown in Figure 4.3b. Near the poles the value of $|H(s)|$ becomes very large and the surface surges upwards. For values of s near the zero at $s = 0$, $|H(s)|$ becomes smaller and the surface dips down to reach zero at the origin. Only the left-half plane values of $|H(s)|$ are shown. We have effectively cut through the surface along the ω-axis of the s-plane. The profile of the surface that we see represents the values of $|H(s)|$ for $s = j\omega$.

You can visualize the magnitude surface of a function $H(s)$ as a kind of 'Big-Top' tent. The poles of $H(s)$ act as tent poles to support the surface while the zeros of $H(s)$ act as pegs to pin the surface down to the ground. The surface surges up at each pole and sags to a lower level elsewhere.

Thus we gain some insight into how the frequency-response function (or impulse-response spectrum), is determined by the associated pole-zero pattern. Near the zero at $s = j\omega = 0$ the value of $|H(j\omega)|$ is small. As we move up the ω axis in the direction of increasing frequency the value of $|H(j\omega)|$ increases, reaching its maximum value as we pass the pole at $s = -1/2 + j\sqrt{3}/2$ on our left. Further increase in frequency takes us away from the pole and the value of $|H(j\omega)|$ falls slowly towards zero at very high frequencies.

To summarize, therefore, zeros tend to reduce the value of the magnitude function $|H(s)|$ while poles tend to increase it. When zeros occur near the

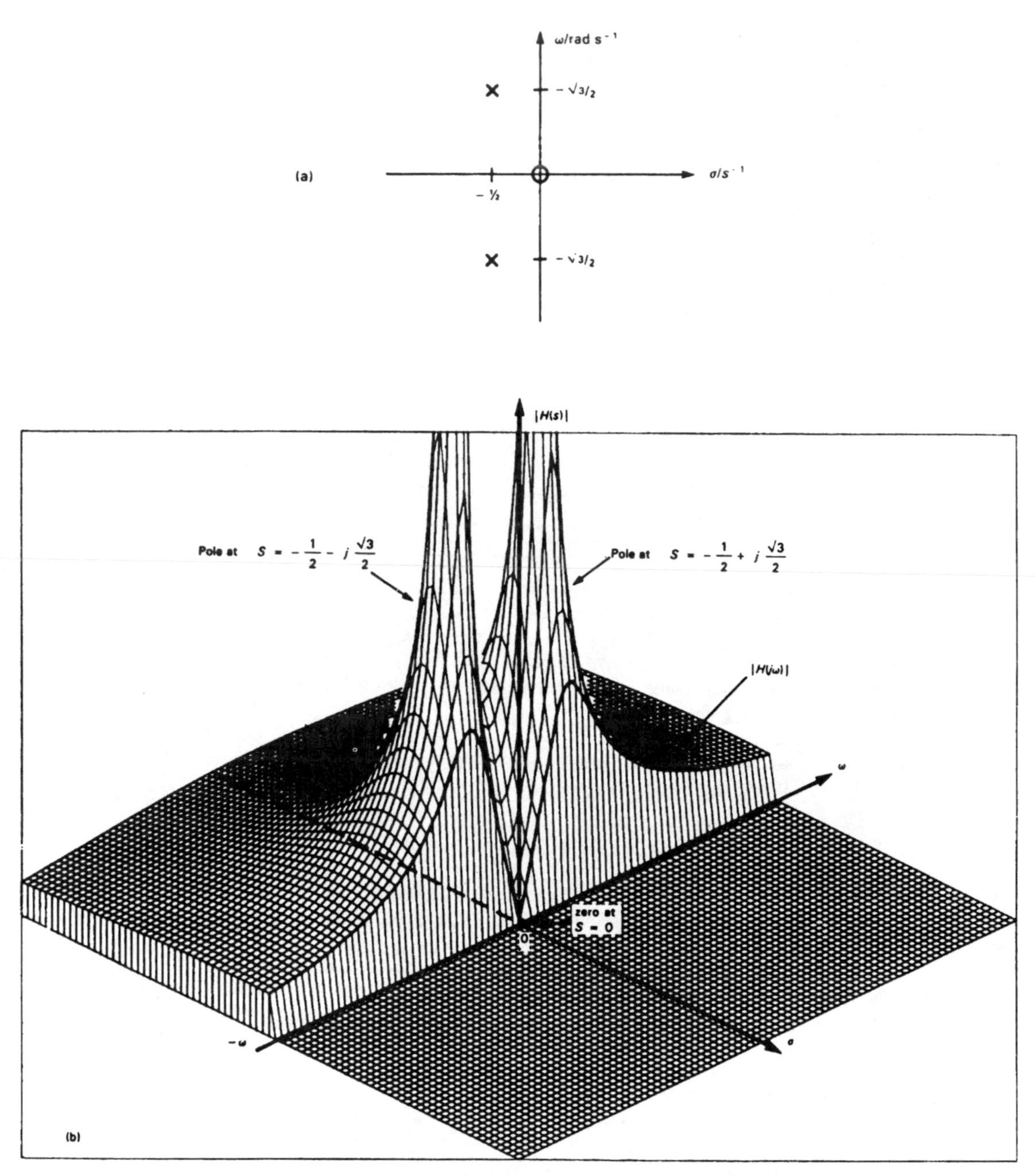

Fig. 4.3 (a) Pole-zero-diagram of $H(s) = s/(s^2 + s + 1)$; 3-D representation of left-half plane magnitude of $H(s)$.

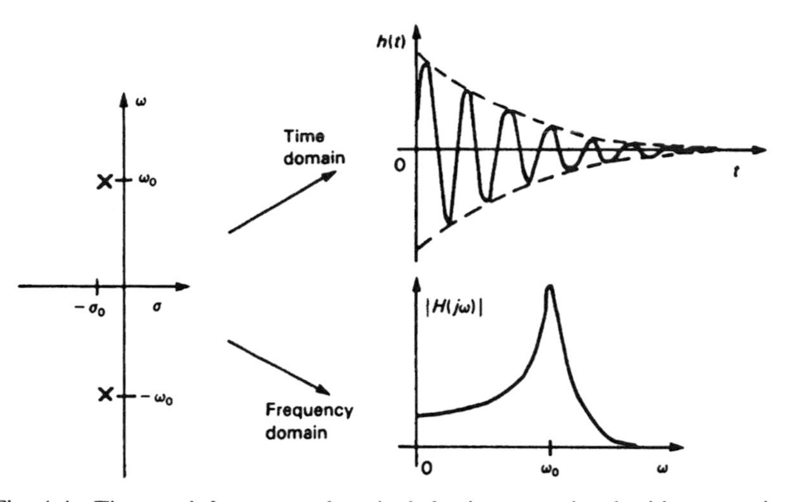

Fig. 4.4 Time and frequency domain behaviour associated with a complex conjugate pair of poles.

imaginary axis in the s-plane where $s = j\omega$, the corresponding magnitude-response function $|H(j\omega)|$ (or spectral magnitude function in the case of a signal), will be reduced at frequencies in the neighbourhood of the zero. If a zero exists on the ω-axis at $s = j\omega_0$ the magnitude function will become zero at the frequency $\omega = \omega_0$. Poles that lie relatively close to the imaginary axis will, in general, cause the magnitude function to peak at frequencies in the neighbourhood of the pole.

Figure 4.4 illustrates how the frequency-domain and time-domain ideas are related for the specific example of a lightly-damped two pole system. The system poles lie close to the imaginary axis in the s-plane at the complex-conjugate positions $s = -\sigma_0 \pm j\omega_0$. The corresponding impulse response of the system is a lightly-damped decaying sinewave of frequency ω_0 and decay factor σ_0, while the frequency magnitude-response function is peaked around ω_0 owing to the nearby pole.

The shape of the magnitude-response function shows that the system is acting as a bandpass filter. Input frequency components around $\omega = \omega_0$ are attentuated least by the system, while the amplitudes of very low and very high frequency components are considerably reduced. $|H(j\omega)|$ approaches zero at high frequencies because $H(s)$ has a zero at infinity.

Summary

The Laplace transform can handle a wider range of signals than can the Fourier transform. The Laplace models of signals and linear systems that we have met have, with one or two exceptions, been rational functions of the complex frequency variable s. By identifying the values of s for which the functions become infinitely large, or zero, we can represent a Laplace model in terms of its poles and zeros in the s-plane.

One of the important features of Laplace transform techniques is that certain time-domain operations become different and often simpler operations in the s-domain. For example, differentiation and integration become respectively multiplication and division by s. This leads to the idea that linear differential equations in the time domain can be transformed to simpler algebraic equations in the s-domain.

Of particular importance is the result that the convolution of two time-domain functions turns into multiplication of the corresponding Laplace functions. If $X(s)$ is the Laplace transform of an input signal and $H(s)$ represents the transfer function of a system, then the transform of the output $Y(s)$ is given by

$$Y(s) = H(s) \cdot X(s)$$

If the input is a unit impulse then $X(s) = 1$ and $Y(s) = H(s)$. That is, the transfer function of a linear system is identical to the transform of the impulse response $h(t)$ of the system.

The transfer function $H(s)$ is a complete description of the dynamic behaviour of a system and can yield information about stability, transient response and steady-state behaviour. With practice, the pole-zero diagram of $H(s)$ can be interpreted to give information about both the frequency-response characteristics and the time-domain transient behaviour of the system.

We can replace s by $j\omega$ in signal transforms to obtain the frequency spectrum of the signal, and in transfer functions to obtain the frequency-response function of the system. We noted that this procedure is valid for causal signals only if the associated Fourier transform of the signal exists, and for systems only when the system is stable.

Problems

4.1 Find the Laplace transforms of the time-domain waveforms

(a) $2e^{-2t} + e^{-t}$ (b) $(t - 3)u(t - 3)$

(c) $\sin(3t + \pi/4)$ (d) $e^{-3t}\cos 4t$

4.2 Find the inverse Laplace transforms of the functions

(a) $\dfrac{s + 1}{s(s + 2)}$ (b) $\dfrac{1 - e^{-s}}{s + 3}$

(c) $\dfrac{1}{s^2 + 2s + 10}$ (d) $\dfrac{1}{(s + 4)^3}$

4.3 In a capacitor C and an inductor L the voltage and current are related by:

(a) $i(t) = C\dfrac{dv(t)}{dt}$ for a capacitor

(b) $v(t) = L\dfrac{di(t)}{dt}$ for an inductor

Find the corresponding relationships in the s-domain if at $t = 0$ the capacitor is charged to a voltage $v(0)$ and the inductor carries an initial current $i(0)$.

4.4 An RC network has the transfer function $H(s) = 1/(1 + sRC)$. Use the Laplace transform method to find the time response of the network to:

(a) A rectangular voltage pulse of height 1 volt and duration 0.1 seconds;
(b) an exponential pulse $v_i(t) = 5e^{-20t}$.

Assume that the RC time-constant of the network is 10 ms.

4.5 If $f(t)$ and $F(s)$ are a Laplace transform pair, prove that replacing s by $s + \alpha$ in the transform is equivalent to multiplication by $e^{-\alpha t}$ in the time domain, that is

$$e^{-\alpha t}f(t) \leftrightarrow F(s + \alpha).$$

Use this s-shift property to work out the inverse transform of $(s + 3)/(s + 2)^2$.

4.6 A system is modelled by the transfer function

$$H(s) = \frac{4s + 1}{s^2 + 4s + 13}.$$

Sketch the pole-zero diagram of $H(s)$ and work out the unit-impulse response of the system.

4.7 Plot the pole-zero diagram for each of the following transfer functions and hence sketch on scaled axes the form of the frequency-response magnitude function $|H(j\omega)|$ associated with each system.

(a) $\dfrac{1}{s^2 + \sqrt{2}s + 1}$ (b) $\dfrac{(s^2 + 1)}{(s + 1)^2}$

(c) $\dfrac{s^2 + 2}{(s + 1)(s^2 + 0.2s + 1)}$ (d) $\dfrac{s^2 - s + 1}{s^2 + s + 1}$

4.8 The voltage transfer function of a closed-loop feedback system has the form

$$\frac{V_o}{V_i}(s) = \frac{KH(s)}{1 - KH(s)}$$

where $H(s) = s/(s^2 + 3s + 1)$ and K is a variable gain. As K varies, the locations of the poles of $V_o(s)/V_i(s)$ change. Sketch on a pole-zero diagram the poles of the closed-loop system for $K = 0, 1, 2, 3, 4$. At what value of K does the closed-loop system become unstable?

4.9 An electrical network is modelled by the differential equation

$$\frac{d^2v(t)}{dt^2} + 3\frac{dv(t)}{dt} + 4v(t) = 5i(t) + \frac{di(t)}{dt}.$$

where $v(t)$ represents the output voltage obtained in response to an input current $i(t)$. Assuming $v(t)$ and $dv(t)/dt$ are zero before $t = 0$, find the transfer function $V(s)/I(s)$ of the network. If the input current varies sinusoidally at a frequency of 1 rad s^{-1} with an amplitude of 1 mA, what will be the corresponding steady state output voltage $v(t)$?

4.10 A system is modelled by the transfer function

$$H(s) = \frac{3}{(s + 2)(s + 3)}.$$

Find an expression that describes the response of the system to an input step of height 2 units. What will be the final value of the step response as $t \to \infty$? Estimate the time for the step response to reach 95% of its final value.

5 z-Transforms

Objectives

- ☐ To introduce the z-transform representation of discrete-time signals and systems.
- ☐ To define the transfer function of a discrete-time system and to show how this is related to the unit-sample response.
- ☐ To look at examples of stable and unstable systems.
- ☐ To show how the transfer function and frequency-response function are related for a discrete-time system.
- ☐ To introduce pole-zero descriptions of discrete-time systems in the complex z-plane.
- ☐ To show how the frequency response and the stability of a system can be inferred from its pole-zero description.

The convolution of $x[n]$ and $h[n]$ was defined in Chapter 2 by the *convolution sum*

$$y[n] = \sum_{k=-\infty}^{\infty} x[k]h[n-k]$$
$$= x[n] * h[n]$$

In this chapter we return again to the discussion of discrete-time signals and systems that we started in Chapter 2. There we established the important result that if we know the unit-sample response of a processor then, in principle, we can always work out the output sequence $y[n]$ produced by the processor in response to an arbitrary input sequence $x[n]$. The output sequence $y[n]$ is related to the input $x[n]$ and the unit-sample response $h[n]$ by the *convolution* of $x[n]$ with $h[n]$.

Although convolution is one of the most fundamental ideas in signal processing, the convolution sum approach is often an inconvenient way of handling input–output relationships. Not only is it impossible in practice to process infinite length sequences exactly because the convolution sum would take an infinite amount of time to compute but, more importantly for our purposes, the convolution sum expression makes it difficult to gain much intuitive insight into the dynamic behaviour of a processor.

The z-transform offers a way around such difficulties. By using an algebraic represention of sequences the z-transform provides a straightforward way of handling finite and infinite length unit-sample responses, and a wide range of commonly used deterministic signal sequences. As we shall see, this algebraic approach leads to a compact pole-zero representation analogous to the s-plane models of continuous signals and systems.

The z-transform

For a general sequence $x[n] = \ldots x[-2], x[-1], x[0], x[1], x[2], \ldots$ the z-transform $X(z)$ of $x[n]$ is defined as

$$X(z) = \sum_{n=-\infty}^{\infty} x[n]z^{-n}. \qquad (5.1)$$

We shall be dealing only with *causal sequences*, for which $x[n] = 0$ for $n < 0$, so the summation can be taken from $n = 0$:

$$X(z) = \sum_{n=0}^{\infty} x[n] z^{-n}$$
$$= x[0] z^{0} + x[1] z^{-1} + x[2] z^{-2} + \ldots + x[n] z^{-n} + \ldots \quad (5.2)$$

We shall not attempt to provide a rigorous mathematical treatment of the z-transform, but will introduce a number of its important features in the context of system behaviour. A comprehensive discussion is to be found in Oppenheim and Schafer, *Discrete-Time Signal Processing*, Prentice-Hall, 1989.

The z-transform is thus a power series in z^{-1} with the coefficients provided by the sequence of sample values $x[n]$. For example, the z-transform of the finite length sequence

$$x[n] = 3, 2, 1, 0, 0, 0, \ldots$$

is simply the sum

$$X(z) = 3z^{0} + 2z^{-1} + z^{-2}$$

which, since $z^{0} = 1$, can be written as

$$X(z) = 3 + 2z^{-1} + z^{-2}.$$

One way of looking at the z-transform of a sequence is to think of z^{-1} as being used to 'tag' or mark the place of a particular sample. In the sequence 3, 2, 1, 0, 0, . . . the sample of value 3 arrives first at $t = 0$. The sample of value 2 arrives next; it is delayed relative to the first sample by one sampling period and is tagged by z^{-1} to indicate this delay. The final sample of value 1 arrives two sampling periods after the first and is tagged by z^{-2} to indicate the fact. The use of z^{-1} helps us to keep the books straight when we are dealing with time sequences of numbers.

Worked Example 5.1

Write down the z-transforms of the sequences shown in Figure 5.1.

Solution:

(a) $\delta[n]$ is the unit-sample sequence 1, 0, 0, 0, 0, . . .
The z-transform is $1z^{0} + 0 + 0 + \ldots = 1$.

(b) $\delta(n - 1)$ is the shifted unit-sample sequence 0, 1, 0, 0, 0, 0, . . .
The z-transform is $0 + 1z^{-1} + 0 + 0 = z^{-1}$.

(c) $h(n)$ is the sequence $\frac{1}{2}, \frac{1}{2}, 0, 0, 0, \ldots$
The z-transform is $\frac{1}{2}z^{0} + \frac{1}{2}z^{-1}$. Since the z-transform obeys the rules of ordinary algebra we can also express this as

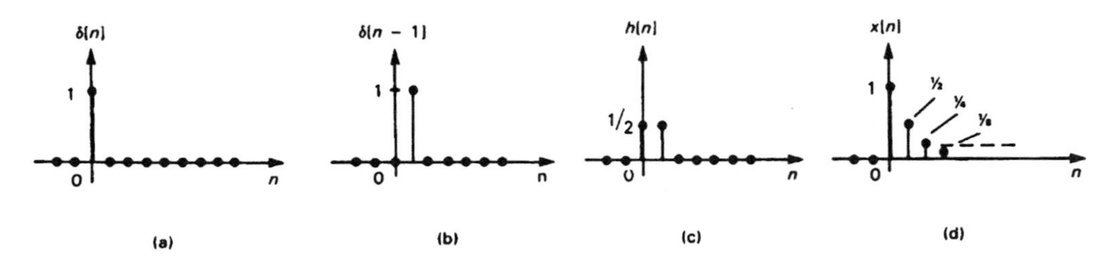

Fig. 5.1

$$H(z) = \frac{1 + z^{-1}}{2}.$$

(d) $x(n)$ is the sequence $1, \frac{1}{2}, \frac{1}{4}, \frac{1}{8}, \ldots, (\frac{1}{2})^n, \ldots$
The z-transform is

$$X(z) = 1 + \tfrac{1}{2}z^{-1} + \tfrac{1}{4}z^{-2} + \tfrac{1}{8}z^{-3} + \ldots + (\tfrac{1}{2})^n z^{-n} + \ldots$$

For causal sample sequences the relationship between a sequence $x[n]$ and its z-transform is unique. No two sequences have the same z-transform (unless they are identical) and, similarly, no two transforms correspond to the same sequence. We denote this one-to-one relationship in the usual way by a double-headed arrow

$$x[n] \leftrightarrow X(z) \tag{5.3}$$

and say that $x[n]$ and $X(z)$ are a *z-transform pair*.

Exercise 5.1 Write down the z-transforms of the sequences:

(a) $x[n] = 1, 1, 1, 0, 0, 0, \ldots$
(b) $x[n] = \alpha^n$, for $0 \leqslant n \leqslant \infty$

Taking the z-transform of a sequence is a linear operation. As Worked Example 5.1 showed, the principles of superposition and homogeneity also apply to z-transforms.

Superposition If we have the transform pairs

$$x[n] \leftrightarrow X(z) \text{ and } y[n] \leftrightarrow Y(z)$$

then the transform of the sum of the sequences is equal to the sum of the transforms:

$$x[n] + y[n] \leftrightarrow X(z) + Y(z). \tag{5.4}$$

Homogeneity If we scale a sequence by a constant factor k, so that

$$kx[n] = kx[0], kx[1], kx[2], kx[3], \ldots$$

then the transform is also scaled by the same factor:

$$kx[n] \leftrightarrow kX(z). \tag{5.5}$$

The z-notation is particularly useful for handling delays. For example, the sequence $3, 2, 1, 0, 0, 0, \ldots$ delayed by one sampling period is

$$0, 3, 2, 1, 0, 0, 0, \ldots$$

and is represented by the z-transform

$$0 + 3z^{-1} + 2z^{-1} + z^{-3}$$

or simply

$$3z^{-1} + 2z^{-2} + z^{-3}$$

Multiplication by z^{-1} denotes a *delay* of one sampling period while multiplication by z denotes an *advance* of one sampling period.

which is just the transform of the original sequence multiplied by z^{-1}. Multiplication by z^{-1}, therefore, denotes a delay of one sampling period. For this reason z^{-1} is sometimes called the *delay operator*.

The delay property of z^{-1} can then be expressed formally as follows:

> If a sequence $x[n]$ is represented by the transform $X(z)$ then the sequence delayed by k sampling periods, $x[n - k]$, has the transform $z^{-k}X(z)$.

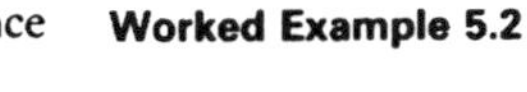

Worked Example 5.2

Write down the z-transforms of the sequences shown in Figure 5.2 and hence show that $X_1(z) = z^{-1}X(z)$ and $X_2(z) = z^{-2}X(z)$.

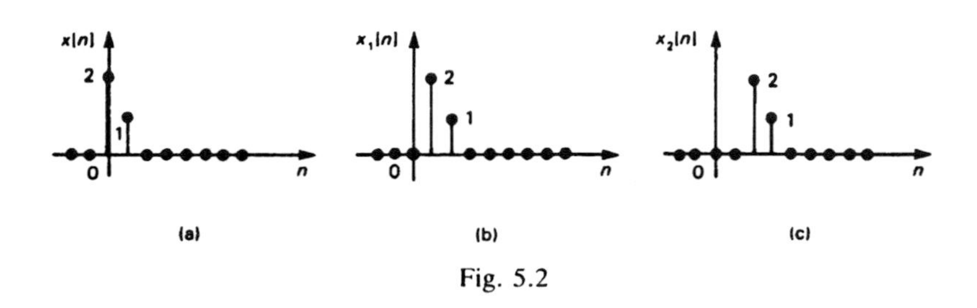

Fig. 5.2

Solution: We note first of all that $x_1[n] = x[n - 1]$ and $x_2[n] = x[n - 2]$. The z-transforms are given by

$$\begin{aligned}
x[n] &\leftrightarrow X(z) = 2z^0 + z^{-1} = 2 + z^{-1}.\\
x[n-1] &\leftrightarrow X_1(z) = 2z^{-1} + z^{-2} = z^{-1}X(z).\\
x[n-2] &\leftrightarrow X_2(z) = 2z^{-2} + z^{-3} = z^{-2}X(z).
\end{aligned}$$

The transfer function

We are now ready to show how the z-transform can be used to model the relationship between the input and ouput sequences of a discrete-time processor. Consider first the moving averager shown in Figure 5.3a, and described by the linear difference equation

$$y[n] = \frac{x[n] + x[n-1]}{2}. \tag{5.6}$$

The first step is to replace each of the sample sequences appearing in this equation by its z-transform. The output sequence $y[n]$ is replaced by $Y(z)$, the input sequence $x[n]$ by $X(z)$ and the shifted input sequence $x[n - 1]$ by $z^{-1}X(z)$ to give

$$Y(z) = \frac{X(z) + z^{-1}X(z)}{2}. \tag{5.7}$$

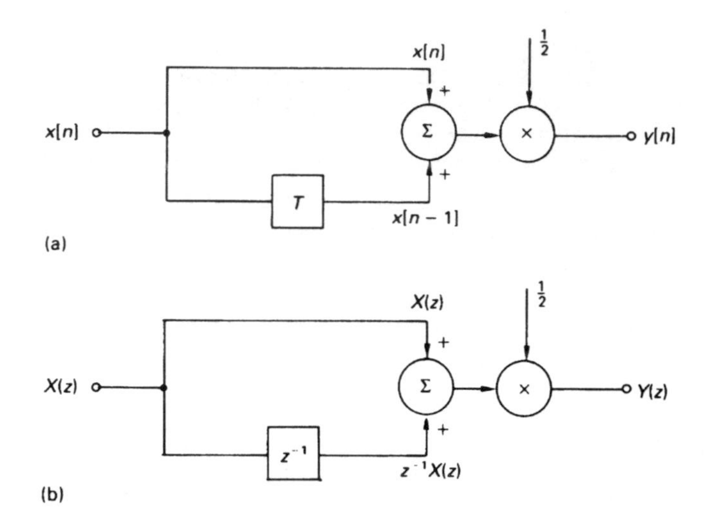

Fig. 5.3 (a) Block diagram of moving averager; (b) z-transform version.

Figure 5.3b shows the block diagram form of this expression. Collecting terms, using the rules of ordinary algebra, gives

$$Y(z) = \frac{1 + z^{-1}}{2} X(z). \tag{5.8}$$

$H(z)$ is defined to be the ratio of the z-transform of the output to the z-transform of the input. The output of the processor is assumed to be zero and unchanging when the input is applied.

If we now divide both sides of this equation by $X(z)$, we get

$$H(z) = \frac{Y(z)}{X(z)} = \frac{1 + z^{-1}}{2}. \tag{5.9}$$

$H(z)$ is called the *transfer function* of the processor. It is usual to express the final form of z-transforms and transfer functions in terms of z rather then z^{-1}. In this case multiplying numerator and denominator by z gives

$$H(z) = \frac{z + 1}{2z}. \tag{5.10}$$

The transfer function $H(z)$ is a property of the *system* and characterises the way the system modifies the input sequence to produce the output sequence. When $H(z)$ has been specified we can find the z-transform of the output sequence for a given input using the relationship

$$Y(z) = H(z) \times X(z). \tag{5.11}$$

A case of particular interest arises when $x[n]$ is defined to be the unit-sample sequence $\delta[n]$. The input will then have the z-transform $X(z) = 1$ and the output $y[n]$ will be the *unit-sample response* $h[n]$ of the system, with the z-transform

$$Y(z) = H(z) \times 1 = H(z). \tag{5.12}$$

This is a very important result, for we have shown that the transfer function of a linear discrete-time processor is equal to the z-transform of the processor's unit sample response. In other words the *unit-sample response* $h[n]$ and the *transfer function* $H(z)$ are a *z-transform pair*:

We have derived analogous results for continuous-time systems involving the Fourier and the Laplace transforms:

$$h(t) \leftrightarrow H(j\omega)$$

$$h(t) \leftrightarrow H(s)$$

$$h[n] \leftrightarrow H(z). \tag{5.13}$$

If, for example, we write the transfer function of the two-term moving averager in the form

$$H(z) = 0.5 + 0.5z^{-1}$$

then Equation 5.13 tells us that we can interpret this expression as the z-transform of the unit-sample response sequence $h[n]$ of the processor, where

$$h[n] = 0.5, 0.5, 0, 0, 0, \ldots$$

Worked Example 5.3

A processor is described by the linear difference equation

$$y[n] = x[n] + 4x[n-2].$$

Find the transfer function of the processor and hence write down the unit-sample response sequence of the processor.

Solution: Replacing $y[n]$ by $Y(z)$, $x[n]$ by $X(z)$ and $x[n-2]$ by $z^{-2}X(z)$ gives the expression

$$Y(z) = X(z) + 4z^{-2}X(z) = (1 + 4z^{-2})X(z).$$

The transfer function is the ratio of the transform of the output to the transform of the input:

$$H(z) = \frac{Y(z)}{X(z)} = 1 + 4z^{-2} = \frac{z^2 + 4}{z^2}.$$

$H(z)$ is equal to the z-transform of the unit-sample response. Writing $H(z)$ out in full gives

$$H(z) = 1z^0 + 0z^{-1} + 4z^{-2}$$

from which the unit-sample response sequence $h[n]$ of the processor is

$$h[n] = 1, 0, 4, 0, 0, 0, \ldots$$

Exercise 5.2

Find the transfer functions of the systems defined by

(a) the linear difference equation $y[n] = x[n-1] + \alpha x[n-2]$;
(b) the unit-sample response $h[n] = 0, 1, 0, -1, 0, 0, 0, \ldots$

System response

If we know the transfer function $H(z)$ of a processor and can work out the z-transform $X(z)$ of the input sequence, then the z-transform $Y(z)$ of the output sequence is given by the product $H(z)X(z)$. Consider again the processor and input sequence we used in Worked Example 2.3. An input sequence

$$x[n] = 3, 1, 2, -1, 0, 0, \ldots$$

is applied to a processor with the unit-sample response

$$h[n] = 1, 2, 1, 0, 0, \ldots$$

In Chapter 2 we used convolution to work out the output sequence $y[n]$ and found that

$$\begin{aligned} y[n] &= x[n] * h[n] \\ &= 3, 7, 7, 4, 0, -1, 0, 0, \ldots \end{aligned}$$

Now we can repeat the calculation using z-transforms. The transforms corresponding to the sequences $x[n]$ and $h[n]$ are:

$$X(z) = 3 + z^{-1} + 2z^{-2} - z^{-3}$$

and

$$H(z) = 1 + 2z^{-1} + z^{-2}.$$

For the moment we shall leave the transforms in terms of z^{-1}. Now, since we know that the transform $H(z)$ of the unit-sample response gives us the transfer function of the processor, we can express the transform $Y(z)$ of the processor's response to the input sequence as

$$\begin{aligned} Y(z) &= H(z) \times X(z) \\ &= (1 + 2z^{-1} + z^{-2}) \times (3 + z^{-1} + 2z^{-2} - z^{-3}). \end{aligned}$$

Using the normal rules of polynomial multiplication, this product can be evaluated as the sum

$$\begin{aligned} Y(z) = (3 + 6z^{-1} + 3z^{-2}) & \\ + (z^{-1} + 2z^{-2} + z^{-3}) & \\ + (2z^{-2} + 4z^{-3} + 2z^{-4}) & \\ - (z^{-3} + 2z^{-4} + z^{-5}). & \end{aligned}$$

If you compare this result with that in Worked Example 2.3, you will see that the four bracketed expressions correspond to the z-transforms of the processor's response to the input sequences $3\delta[n]$, $\delta[n - 1]$, $2\delta[n - 2]$ and $-\delta[n - 3]$ respectively.

If we now add these expressions, taking care to combine only the coefficients for the same powers of z^{-1}, we get

$$Y(z) = 3 + 7z^{-1} + 7z^{-2} + 4z^{-3} + 0z^{-4} - 1z^{-5}$$

which you should be able to recognise as the z-transform of the sequence

$$y[n] = 3, 7, 7, 4, 0, -1, 0, 0, \ldots$$

The rules for polynomial multiplication, which involve scalar multiplication, shifting and adding, are the same as those for convolution.

This result is the same as we found in Worked Example 2.3. However, what is important here is that we have reached this result by *multiplication* of $X(z)$ and $H(z)$ rather than by the *convolution* of $x[n]$ and $h[n]$. Expressing sequences and transfer functions as algebraic functions of z^{-1} or z, therefore, leads us naturally to a procedure where multiplication rather than convolution is used to relate the input, output and unit-sample sequences of a linear discrete-time processor.

Exercise 5.3

An input sequence

$$x[n] = 1, -1, 1, 0, 0, 0, \ldots$$

is applied to a processor with a unit-sample response

$$h[n] = 1, 2, 1, 0, 0, 0, \ldots$$

Use polynomial multiplication to find the processor's output sequence.

Infinite sequences

So far we have been concerned only with the z-transforms of finite-length sequences. As you saw in Chapter 2, however, infinite sequences are not uncommon. How can we find the z-transforms of these? An intuitive approach is to use the idea that the causal sequences associated with a wide range of common signal models can be thought of as the unit-impulse responses of linear discrete-time processors. Since the unit impulse response $h[n]$ and the transfer function $H(z)$ of a processor are a z-transform pair, the z-transform of the causal sequence is identical to the transfer function of the processor.

To illustrate this idea consider the simple recursive processor shown in Figure 5.4. The processor is described by the linear difference equation

$$y[n] = x[n] + \alpha y[n-1]. \tag{5.14}$$

For $\alpha = 0.5$ this is the processor described in Chapter 2.

To find the transfer function we replace the input $x[n]$ by $X(z)$ and the delayed output $y[n-1]$ by $z^{-1}Y(z)$. The transform equation is

$$Y(z) = X(z) + \alpha z^{-1}Y(z). \tag{5.15}$$

Collecting input and output terms we obtain

$$Y(z) - \alpha z^{-1}Y(z) = X(z) \tag{5.16}$$

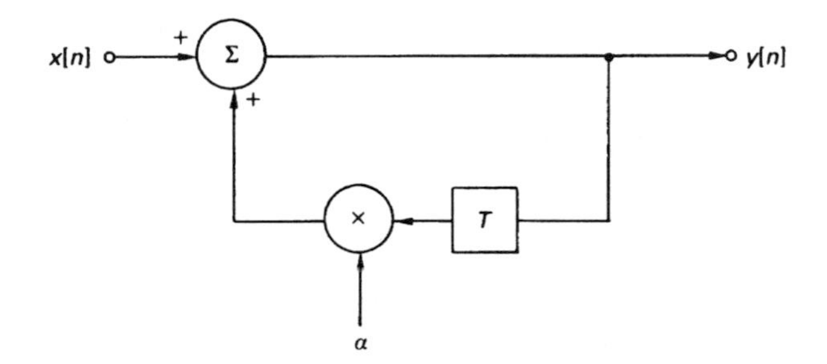

Fig. 5.4 Recursive processor with $\alpha = 0.5$.

or

$$(1 - \alpha z^{-1})\,Y(z) = X(z). \tag{5.17}$$

Therefore the transfer function is

$$H(z) = \frac{Y(z)}{X(z)} = \frac{1}{1 - \alpha z^{-1}} = \frac{z}{z - \alpha}. \tag{5.18}$$

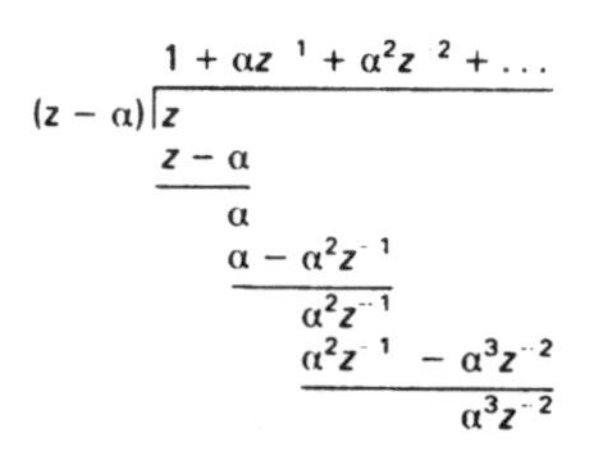

This result is quite unlike any of the transforms we have considered so far, and it is not immediately obvious how we can associate this function with a sample sequence corresponding to the unit-sample response $h[n]$ of the recursive processor. One way to proceed is to express $H(z)$ as a power series in z^{-1} by a process of long division to obtain

$$H(z) = 1 + \alpha z^{-1} + \alpha^2 z^{-2} + \alpha^3 z^{-3} + \ldots \tag{5.19}$$

which we recognise immediately as the z-transform of the infinite sequence

$$h[n] = 1, \alpha, \alpha^2, \alpha^3, \alpha^4, \ldots \tag{5.20}$$

The mathematical conditions for the *z*-transform of an infinite sequence to exist depend upon the convergence properties of the power series expansion. Discussion of this topic is beyond the scope of this book; for more details see Oppenheim and Schafer, *Discrete-Time Signal Processing*, Prentice Hall, 1989 or Papoulis, A., *Circuits and Systems – A Modern Approach*, Holt-Saunders, 1980.

Because the relationship between a sequence and its z-transform is unique, we can now establish a useful z-transform pair. Since Equation 5.19 is simply an expanded version of the expression $z/(z - \alpha)$ we have

$$1, \alpha, \alpha^2, \alpha^3, \alpha^4, \ldots \leftrightarrow \frac{z}{z - \alpha}. \tag{5.21}$$

Here we have been able to express the z-transform of an infinite sequence as a rational function of z. Subject to certain conditions being met it turns out that we are able to do this for many of the infinite sample sequences that we come across, particularly where we know the mathematical form of the sampled signal.

Consider, for example, a waveform modelled as a decaying exponential function $x(t) = \mathrm{e}^{-t}$. What is the z-transform of the sampled version of this waveform? Figure 5.5 shows the waveform sampled every T seconds. The value of $x(t)$ at time $t = 0$ is 1, so this is the first sample value. At the second sampling instant $t = T$ the value of $x(t)$ has fallen to e^{-T}, so this is the value of the second sample. Repeating this process indefinitely at intervals of T gives the infinite sequence of samples:

$$x[n] = 1, \mathrm{e}^{-T}, \mathrm{e}^{-2T}, \mathrm{e}^{-3T}, \mathrm{e}^{-4T}, \ldots \tag{5.22}$$

which can also be written in the form

$$x[n] = 1, \mathrm{e}^{-T}, (\mathrm{e}^{-T})^2, (\mathrm{e}^{-T})^3, (\mathrm{e}^{-T})^4, \ldots \tag{5.23}$$

If now we replace e^{-T} by α we obtain the sequence in Equation 5.20. Then, using the relationship in Equation 5.21, we can write down the z-transform of the sampled exponential waveform:

$$X(z) = \frac{z}{z - \mathrm{e}^{-T}}. \tag{5.24}$$

Note that the transform of the sampled waveform includes the sampling period T explicitly. This reflects the fact that the value of T will affect the values of the individual samples in the sequence.

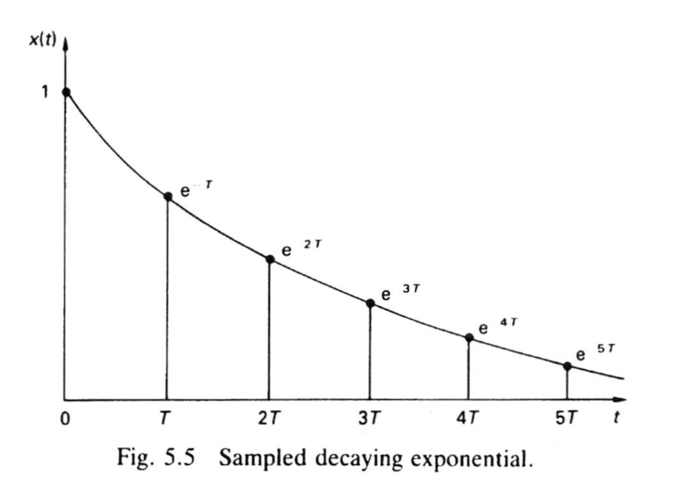

Fig. 5.5 Sampled decaying exponential.

Worked Example 5.4

A unit-step function $u(t)$ is sampled every T seconds. Find the z-transform of the resulting sample sequence, assuming that the value of the unit step at time $t = 0$ is 1. What is the z-transform of the sampled unit step delayed by T seconds?

Solution: The unit-step function is defined as

$$u(t) = 1 \; (t \geqslant 0)$$
$$= 0 \; (t < 0)$$

so the sampled step is the infinite sequence of unit samples

$$u[n] = 1, 1, 1, 1, \ldots$$

Using the transform pair in Equation 5.21 for $\alpha = 1$ gives the z-transform of the sampled unit step as

$$U(z) = \frac{z}{z - 1}.$$

If the sequence is delayed by one sampling interval then its transform is multiplied by z^{-1}. The z-transform of the delayed sampled step is therefore

$$z^{-1}U(z) = \frac{z^{-1}z}{z - 1} = \frac{1}{z - 1}.$$

In practice, the z-transforms of many common sample sequences have been worked out so that it is rarely necessary to evaluate a transform from first principles. A short table of some of the more common z-transform pairs is given in the Appendix.

Exercise 5.4 Use the table in the Appendix to find the z-transforms of the following sampled waveforms:

(a) $3e^{-2t}$, sampled at 1 second intervals;
(b) $2\cos 3t$, sampled at 0.1 second intervals;
(c) $1 - e^{-4t}$, sampled at 0.5 second intervals;
(d) $3e^{-t}$, sampled at 2 second intervals.

Assume that in each case the waveform is zero before $t = 0$.

Notice that the sequences obtained by sampling waveforms (a) and (d) are identical. Thus the associated z-transforms will be identical. This is because a z-transform defines a *sequence* and not the underlying continuous waveform. Since identical sequences can clearly arise from different waveforms sampled at different rates, a sample sequence alone is not sufficient to define uniquely the form of the original waveform.

Just as with Laplace transforms, tables of z-transforms usually include only the basic transform pairs from which more complex functions may be built up. To decompose a function of z the procedure is exactly the same as for Laplace transforms. We first use the partial-fraction technique to express the transform as the sum of simpler elements. Then we look up the sequence corresponding to each partial-fraction term in the table and combine them to give the sequence corresponding to the original transform. The following example shows how this approach is used to work out the step response of a discrete-time processor.

If $X(z)$ is the z-transform of the sequence $x[n]$, then $x[n]$ is called the *inverse* of $X(z)$.

Worked Example 5.5 A recursive discrete-time processor has the transfer function

$$H(z) = \frac{3z}{z - 0.4}.$$

What is the response of the processor to a unit-step input sequence?

Solution: From the table given in the Appendix the z-transform of a unit-step sequence is

$$X(z) = \frac{z}{z - 1}.$$

Hence the z-transform of the response of the processor is given by

$$\begin{aligned} Y(z) &= H(z)X(z) \\ &= \frac{3z}{z - 0.4} \times \frac{z}{z - 1} \\ &= \frac{3z^2}{(z - 0.4)(z - 1)}. \end{aligned}$$

This z-transform is not listed explicitly in the table in the Appendix but we can express $Y(z)$ in partial-fraction form as the sum of two terms and then add the sequences corresponding to each term to find the overall output sequence. $Y(z)$ can be written as

$$\frac{3z^2}{(z - 0.4)(z - 1)} = \frac{Az}{z - 0.4} + \frac{Bz}{z - 1}$$

where A and B are constants. Multiplying both sides of the expression by $(z - 0.4)(z - 1)$ gives

$$3z^2 = Az(z - 1) + Bz(z - 0.4).$$

Putting $z = 1$ eliminates the term containing A, leaving $3 = 0.6B$
Hence $B = 5$. Similarly, putting $z = 0.4$ gives $A = -2$. Thus the partial-fraction expansion of $Y(z)$ is

$$Y(z) = \frac{-2z}{z - 0.4} + \frac{5z}{z - 1}.$$

Using the table of z-transform pairs in the Appendix we find that

$$\frac{-2z}{z - 0.4} \leftrightarrow -2(0.4)^n$$

and

$$\frac{5z}{z - 1} \leftrightarrow 5u[n].$$

So the response of the processor is given by the sum of two sequences

$$y[n] = 5u[n] - 2(0.4)^n.$$

Expanding the sequences term by term gives

$$y[n] = (5, 5, 5, 5, 5, \ldots) - (2, 0.8, 0.32, 0.128, 0.0512, \ldots)$$

from which the output sequence is

$$y[n] = 3, 4.2, 4.68, 4.872, 4.9488, \ldots$$

Notice that as n gets larger the value of $(0.4)^n$ gets smaller. As n increases, therefore, the values of the samples in the output sequence approach 5.

Exercise 5.5

Find the sequences associated with the z-transforms:

(a) $\dfrac{3z}{(z - 0.5)^2}$

(b) $\dfrac{z}{(z - 1)(z - 0.6)}$.

Note that the z-transform of a sequence can usually only be expressed in a compact rational-function form when the sequence is derived from a well-defined deterministic signal model such as a step, a sinusoid or a decaying exponential. An arbitrary sequence derived from sampling a randomly varying voltage (such as a speech waveform in a communication system, for example) will not usually be expressible in this form.

Pole-zero models

In Chapter 4 we showed how the behaviour of a continuous-time system could be inferred from the locations of its poles and zeros in the s-plane. The advantage of this approach was that we could interpret the pole-zero diagram to gain information about the steady-state frequency response, the transient behaviour and the stability of a system. We shall now describe a similar technique for representing the behaviour of discrete-time systems.

By introducing z we have been able to develop models of sequences and discrete-time systems in the form of a ratio of polynomials. As with functions of the Laplace variable s, the values of z for which the numerator polynomial is zero are known as *zeros*, and the values for which the denominator polynomial is zero are known as *poles*. Like s, z is a complex variable with a real and an imaginary part, and we can plot the positions of the poles and zeros at specific values of z on an Argand diagram, just as for s. Not surprisingly, this diagram is known as the *z-plane*.

Worked Example 5.6 Plot the z-plane pole-zero diagrams for the following transfer functions

(a) $H_1(z) = \dfrac{z + 1}{z}$

(b) $H_2(z) = \dfrac{z}{z - \alpha}$

(c) $H_3(z) = \dfrac{z(z + 1)}{z^2 - z + 0.5}$.

Solution:

(a) $H_1(z)$ has zero where $z + 1 = 0$, that is at $z = -1$, and a pole at $z = 0$. Figure 5.6a shows the pole-zero plot.

(b) $H_2(z)$ has a zero at $z = 0$ and a pole at $z = \alpha$, as shown in Figure 5.7b.

(c) $H_3(z)$ has zeros at $z = 0$ and $z = -1$. The pole locations are the values of z which satisfy the equation $z^2 - z + 0.5 = 0$. The solutions are given by

$$z = \frac{1 \pm \sqrt{(1 - 2)}}{2} = 0.5 \pm j0.5$$

Figure 5.6c shows the pole-location diagram with the poles in the complex conjugate positions $z = 0.5 + j0.5$ and $z = 0.5 - j0.5$.

Time-domain interpretation of the z-plane

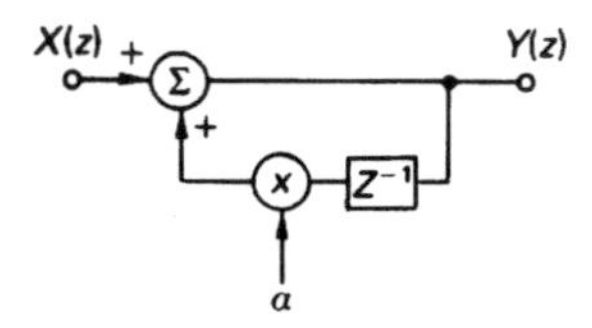

H(z) can be realized as a simple recursive processor.

A z-plane pole-zero diagram is a representation of a particular sequence. If the diagram shows the poles and zeros of the transfer function $H(z)$ of a discrete-time system, then the sequence represented is the unit-sample response of the system. As an example consider again the recursive processor defined by the transfer function

$$H(z) = \frac{z}{z - \alpha}$$

In this case the zero is fixed at $z = 0$ but we can vary the pole position by varying the constant α. Now, we know from Equation 5.21 that the unit-sample response of this system is the infinite sequence

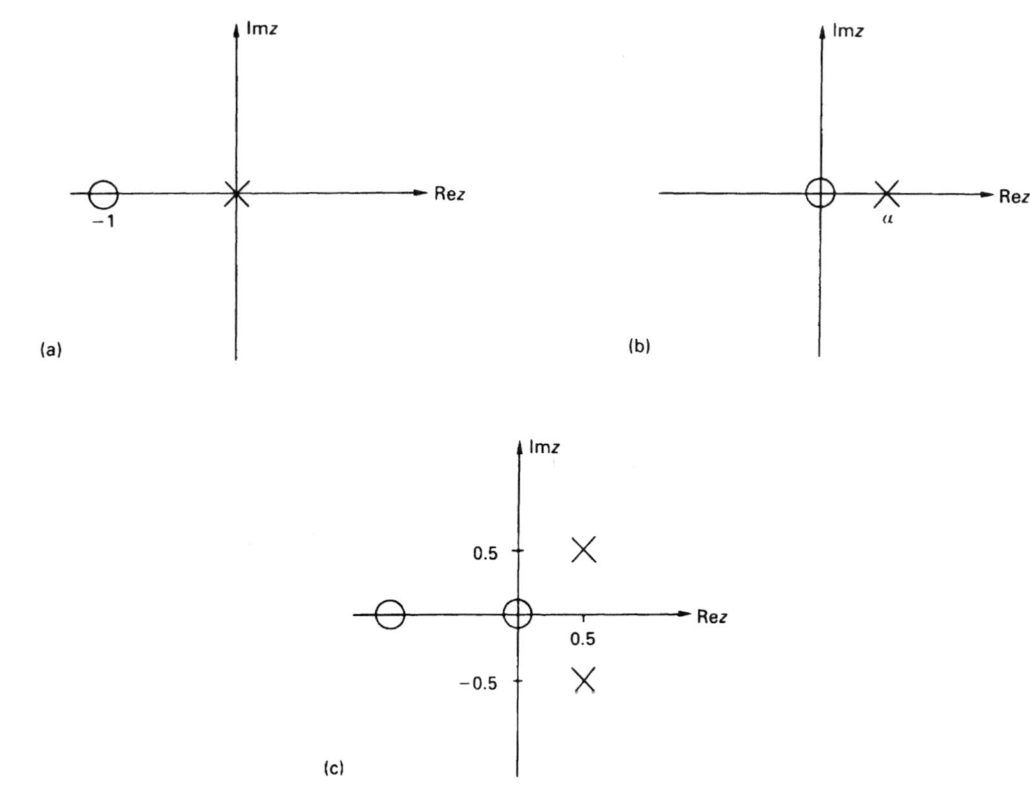

Fig. 5.6 Pole-zero plots of (a) $(z + 1)/z$; (b) $z/(z - \alpha)$; (c) $z(z + 1)/(z^2 - z + 0.5)$.

$$h[n] = 1, \alpha, \alpha^2, \alpha^3, \alpha^4, \ldots$$

so we can easily associate the position of the pole with its corresponding sequence for different values of α.

Figure 5.7 shows the sequences we obtain for a range of values of α. Poles that lie on the real axis between $z = -1$ and $z = +1$ clearly correspond to infinite sequences that die away over successive sampling intervals and indicate that for $-1 < \alpha < 1$, the system is *stable*. Poles that lie beyond $z = +1$ or $z = -1$, on the other hand, are associated with sequences that increase indefinitely and without limit, indicating *unstable* behaviour.

The case where α is equal to 1 or -1 gives rise to poles at $z = +1$ or $z = -1$. This corresponds to behaviour which is on the boundary between stability and instability, since the associated sequences are 1, 1, 1, 1, . . . and 1, -1, 1, -1, . . . which neither grow nor die away. Poles on the real-axis of the z-plane, therefore, correspond to strictly stable systems, or to sequences that die away, only if they lie between $z = -1$ and $z = +1$.

A final point to note is that poles that lie to the left of the origin of the z-plane are associated with infinite sequences of alternating sign. This is a

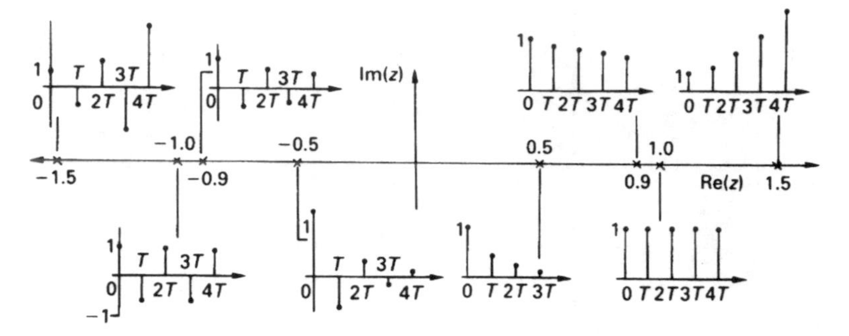

Fig. 5.7 Sequences associated with poles on the real-axis of the z-plane.

direct consequence of putting a negative value of α in the sequence $1, \alpha, \alpha^2, \alpha^3, \alpha^4, \ldots$ Such sequences can be viewed as oscillating with a frequency equal to half the sampling frequency since the sign of the samples changes from positive to negative and back again in two sampling periods.

Worked Example 5.7 Figures 5.8a and 5.8b show block diagrams of two discrete-time processors. The linear difference equations of the processors are

(a) $y[n] = ax[n] + bx[n-1] + cx[n-2]$
(b) $3y[n] = 3x[n] + 7y[n-1] - 2y[n-2]$.

Find the transfer functions of these processors and hence comment on the form of their unit-sample response.

Solution: Figure 5.8a shows a non-recursive system. Replacing the input, output and delayed sequences by their z-transforms gives

$$Y(z) = aX(z) + bz^{-1}X(z) + cz^{-2}X(z)$$

so the transfer function $Y(z)/X(z)$ is

$$H(z) = a + bz^{-1} + cz^{-2}$$
$$= \frac{az^2 + bz + c}{z^2}.$$

This is a finite impulse response (FIR) system. All FIR systems are said to be *unconditionally stable* since, by definition, their unit-sample response takes zero values after a finite number of terms.

$H(z)$ has two poles and two zeros. Both poles lie at $z = 0$ but the locations of the zeros depend upon the values of a, b and c. Since the poles lie between $z = -1$ and $z = +1$, the system will be stable for all finite values of a, b and c. The unit-sample response is simply the three-term sequence $a, b, c, 0, 0, 0, \ldots$

Figure 5.8b shows a recursive processor. We have

$$3Y(z) = 3X(z) + 7z^{-1}Y(z) - 2z^{-2}Y(z).$$

Rearranging and collecting terms gives the transfer function $H(z) = Y(z)/X(z)$, where

$$H(z) = \frac{3}{3 - 7z^{-1} + 2z^{-2}} = \frac{3z^2}{3z^2 - 7z + 2}.$$

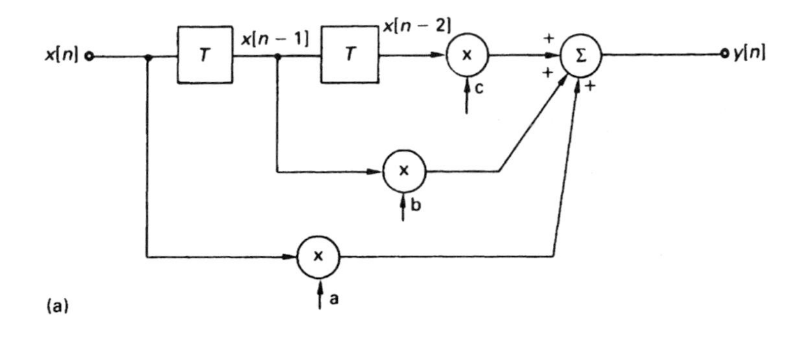

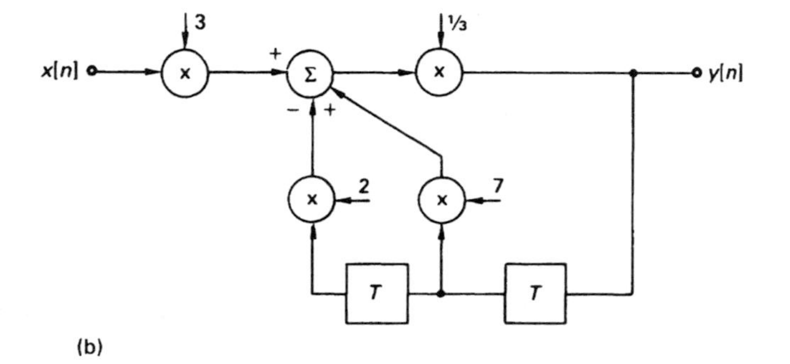

Fig. 5.8 Block diagrams of the processors defined by (a) $y[n] = ax[n] + bx[n-1] + cx[n-2]$; (b) $3y[n] = 3x[n] + 7y[n-1] - 2y[n-2]$.

The denominator of $H(z)$ can be factorized as

$$3z^2 - 7z + 2 = (3z - 1)(z - 2).$$

Hence $H(z)$ has poles at $z = \frac{1}{3}$ and $z = 2$. $H(z)$ can be expressed in terms of partial fractions to indicate the contribution to the unit-sample response from each pole term:

$$H(z) = \frac{Az}{z - \frac{1}{3}} + \frac{Bz}{z - 2}$$

where A and B are constants. The pole at $z = \frac{1}{3}$ lies in the stable region $-1 < z < 1$ and corresponds to a decaying sequence. The pole at $z = 2$, however, lies in the unstable region beyond $z = 1$ and is associated with a response sequence that grows without limit. This is sufficient to establish that the unit-sample response sequence of the system will itself grow without limit, indicating that the system as a whole is *unstable*.

All poles must be associated with stable sequences if the system as a whole is to be stable.

Exercise 5.6

Plot the pole-zero diagrams of the following transfer functions and hence identify which represent stable systems:

(a) $\dfrac{z}{(3z-1)^2}$

(b) $\dfrac{(z-2)(4z^2+z+1)}{z^3}$

(c) $\dfrac{z+1}{z^2+2z+2}$

(d) $\dfrac{z^2}{z^2+2z+0.75}$.

In Worked Example 5.7 we derived the transfer functions of two processors from their linear difference equations. If we start with a transfer function, however, it is straightforward to get back to the defining difference equation. This is useful if we need to use the difference equation in a computer program to calculate the response of a system to a given input sequence. Suppose, for example, the transfer function of a system is

$$H(z) = \frac{Y(z)}{X(z)} = \frac{z^2}{(z^2 - z + 0.5)}.$$

First, we express $H(z)$ in terms of z^{-1} by dividing the numerator and denominator by z^2:

$$H(z) = \frac{Y(z)}{X(z)} = \frac{1}{1 - z^{-1} + 0.5z^{-2}}.$$

Cross-multiplying gives

$$Y(z) - z^{-1}Y(z) + 0.5z^{-2}Y(z) = X(z).$$

Finally, if we replace the transforms $Y(z)$, $z^{-1}Y(z)$ and $z^{-2}Y(z)$ by the corresponding sequences $y[n]$, $y[n-1]$ and $y[n-2]$ we form the associated linear difference equation

$$y[n] - y[n-1] + 0.5y[n-2] = x[n]$$

which can be rearranged to give the output sequence $y[n]$ in terms of the input sequence $x[n]$ and the delayed output sequences $y[n-1]$ and $y[n-2]$:

$$y[n] = x[n] + y[n-1] - 0.5y[n-2].$$

If we know the form of $x[n]$ we can use this equation step by step to work out $y[n]$, just as we did in Chapter 2.

Exercise 5.7 Find the linear difference equations associated with the transfer functions:

(a) $H(z) = \dfrac{z}{(z-0.2)^2}$

(b) $H(z) = \dfrac{z+1}{z-1}$.

A general interpretation of the z-plane

So far we have shown that poles that lie between $z = -1$ and $z = +1$ on the real axis of the z-plane are associated with stable systems or decaying sample sequences. Poles that lie beyond $z = 1$ or $z = -1$, on the other hand, are associated with unstable systems or sequences that grow indefinitely. As we saw in Worked Example 5.6, however, poles are not constrained to lie only on the real axis but can occur in complex-conjugate pairs in the z-plane. The next step is to investigate the general form of the sequences associated with complex z-plane poles, and hence establish a general stability criterion for discrete-time systems.

If a system has complex poles then these will lie in complex-conjugate pairs in the z-plane. Figure 5.9 shows some examples of typical sequences associated with different complex-conjugate pole locations. The important feature to notice is that poles located within a circle of unit radius centred on the origin of the z-plane correspond to sequences whose envelopes (indicated by broken lines) die away with time. The closer the poles are to the origin, the faster this decay takes place (for a given sampling frequency). Poles that lie outside the circle, as in Figure 5.9c, correspond to sequences that increase indefinitely with time, indicating unstable behaviour.

It turns out that the unit-sample response of any system with real or complex-conjugate poles will eventually die away to zero provided that *all* the poles lie within a circle of unit radius centred on the origin of the z-plane. This circle, called the *unit circle*, is usually included on a z-plane pole-zero plot to indicate the region of stability. Clearly the unit circle, interpreted as the boundary between stability and instability, plays the same role in the z-plane as the ω-axis does in the s-plane. This link is further reinforced in the next section where we discuss the frequency response of a discrete-time system.

Frequency response of a discrete-time system

We know that if we replace s by $j\omega$ in the transfer function $H(s)$ of a continuous system we obtain the system's frequency-response function $H(j\omega)$. We shall now extend this technique to discrete-time systems, and show that replacing z by $e^{j\omega T}$ in a transfer function $H(z)$ gives the corresponding frequency-response function of the discrete-time processor.

We begin by considering the steady-state response of a discrete-time processor to a sampled sinusoidal signal of amplitude A and frequency ω, defined by the infinite sequence

$$x[n] = x(nT) = A\cos n\omega T. \qquad (5.25)$$

By using an infinitely long sequence we are able to assume that the input has been present for a very long time, so that any transient effects in the output caused by suddenly applying the input have become small enough to neglect. What we are considering here, therefore, is the *steady-state* behaviour of the system.

Figure 5.10 illustrates the effect of a typical linear processor on the sinusoidal sequence $x[n]$. The steady-state response is a sinusoidal sequence of the same frequency as the input, but with a different amplitude and phase. The output can be modelled as the sequence

$$y[n] = B\cos(n\omega T + \Phi). \qquad (5.26)$$

By analogy with the continuous-time case we can say that the amplitude ratio B/A defines the magnitude of the system frequency response, and Φ defines its phase.

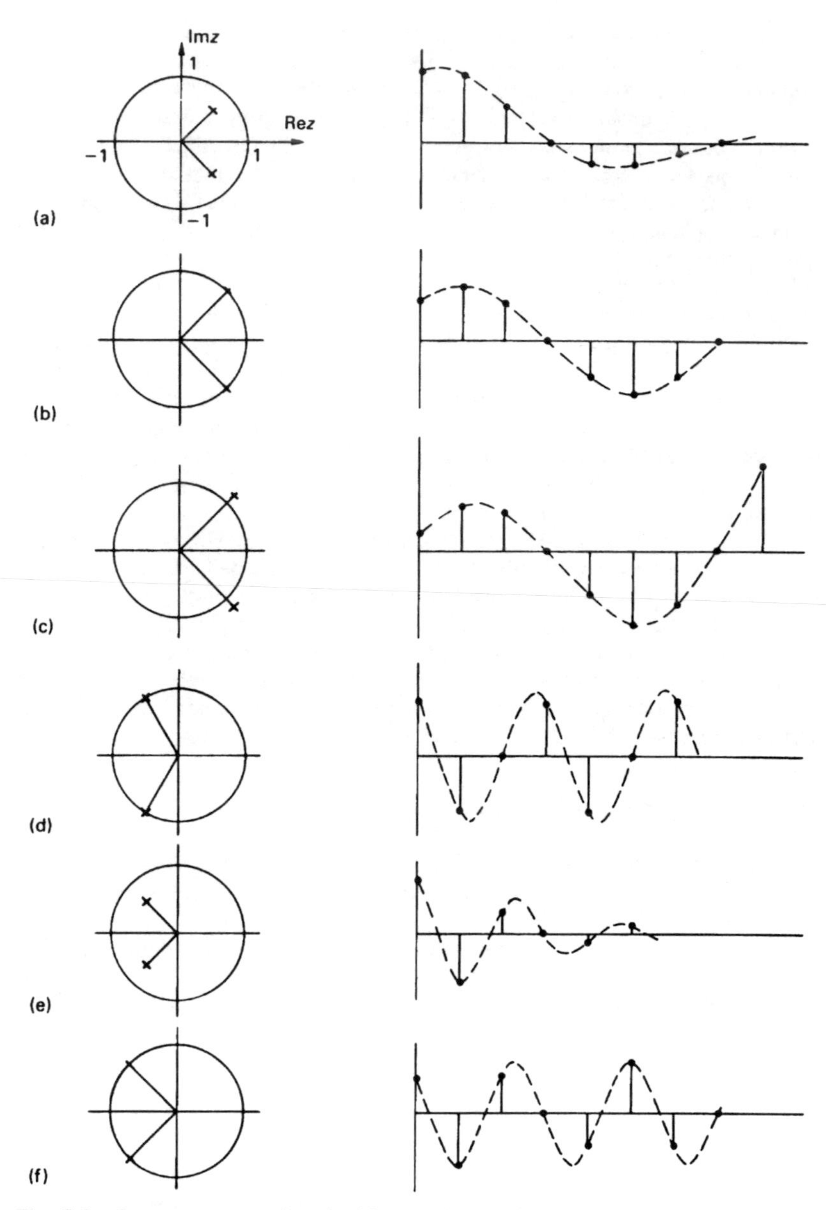

Fig. 5.9 Sequences associated with complex-conjugate pairs of z-plane poles.

To find the form of the frequency response we shall use a technique suggested by our treatment in Chapter 3 of continuous-time signals, and replace the sinusoidal input $x[n]$ by a single, representative, complex exponential component. In this case the input to the discrete-time processor will be the sequence of complex values defined by

$$w[n] = e^{jn\omega T}. \tag{5.27}$$

Now, we know that the response $y[n]$ of a linear system to a general input $x[n]$ can be found from the convolution sum:

$$y[n] = \sum_{k=-\infty}^{\infty} h[k]x[n-k] \tag{5.28}$$

where $h[n]$ is the unit-sample response of the system. Since the input $x[n]$ is the complex exponential sequence $w[n]$, we can write the shifted input sequence as

$$x[n-k] = w[n-k] = e^{j\omega(n-k)T}$$
$$= e^{jn\omega T}e^{-jk\omega T}. \tag{5.29}$$

Substituting for $x[n-k]$ in the convolution sum gives the system response

$$y[n] = \sum_{k=-\infty}^{\infty} h[k]e^{jn\omega T}e^{-jk\omega T}. \tag{5.30}$$

Note that the variable in the convolution sum is k. Since the term $e^{jn\omega T}$ does not vary with k it can be taken outside of the summation

$$y[n] = e^{jn\omega T}\sum_{k=-\infty}^{\infty} h[k]e^{-jk\omega T}. \tag{5.31}$$

This expression gives us all the information we need to determine the effect of a given system on a complex exponential series. We see that the output sequence $y[n]$ is given by the product of the input sequence $x[n] = e^{j\omega nT}$ and the result of a summation over k, from $-\infty$ to ∞. Now, the summation involves the terms of the unit-impulse response $h[n]$ and, for a causal system, $h[n] = 0$ for $n < 0$. Hence $y[n] = 0$ for $n < 0$ and so the summation can be taken from $k = 0$ without changing the result:

$$y[n] = e^{jn\omega T}\sum_{k=0}^{\infty} h[k]e^{-jk\omega T}. \tag{5.32}$$

Notice that the terms in the summation depend on k and ω but not on n. This means that at any given frequency $\omega = \omega_0$ and sampling period $T = T_0$ the result of the summation will be a complex number which acts as a multiplier on all the terms in the input sequence. We can express this more clearly by writing the output sequence as

$$y[n] = e^{jn\omega T}H(e^{j\omega T}) \tag{5.33}$$

where

$$H(e^{j\omega T}) = \sum_{k=0}^{\infty} h[k]e^{-jk\omega T}$$
$$= h[0] + h[1]e^{-j\omega T} + h[2]e^{-j2\omega T} + h[3]e^{-j3\omega T} + \ldots \tag{5.34}$$

The complex exponential sequence $w[n]$ is simply a series of complex numbers with values given by

$$w[n] = e^{jn\omega T}$$
$$= \cos n\omega T + j\sin n\omega T.$$

A sinusoidal sequence can then be expressed as the sum of two complex-conjugate exponential sequences:

$$A\cos n\omega T$$
$$= \frac{A}{2}e^{jn\omega T} + \frac{A}{2}e^{-jn\omega T}$$
$$= \frac{A}{2}e^{jn\omega T} + \frac{A}{2}e^{-jn\omega T}$$
$$= \frac{A}{2}w[n] + \frac{A}{2}w^*[n].$$

Given the system response to $w[n]$, therefore, we can use the principle of superposition to find the response to any input expressed as the sum of complex exponentials.

This is analogous to the result we derived in Chapter 3 for a continuous system. We showed that

$$y(t) = e^{j\omega t}H(j\omega)$$

where the frequency response function $H(j\omega)$ is the Fourier transform of the impulse response $h(t)$.

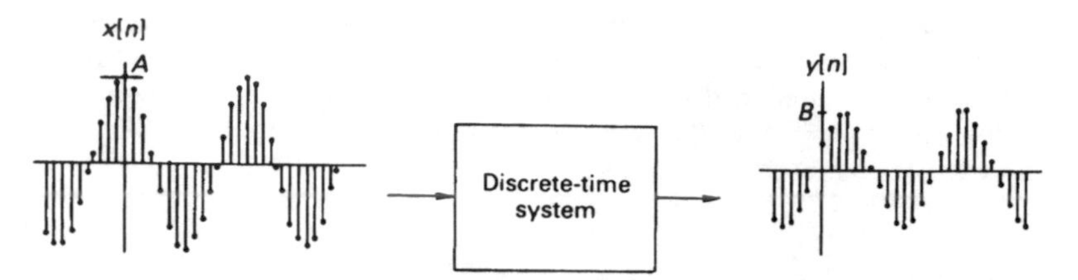

Fig. 5.10 Effect of a discrete-time system on the samples of a sinusoidal signal.

For a given processor the value of $H(e^{j\omega T})$ depends only on the value of the product ωT and the sample values of the unit-sample response $h[n]$. In general, $H(e^{j\omega_0 T_0})$ is a complex number which describes the effect of a discrete-time processor on the sequence obtained by sampling a complex exponential component of frequency $\omega = \omega_0$ at intervals $T = T_0$.

In view of the results obtained in Chapter 3 for the continuous-time case, we shall interpret $H(e^{j\omega T})$ as the *frequency response* of the discrete-time system.

It is now straightforward to make the link between the frequency-response function and the transfer function of a discrete-time processor. We know that the transfer function is equal to the z-transform of the unit-sample response:

$$H(z) = h[0] + h[1]z^{-1} + h[2]z^{-2} + h[3]z^{-3} + \dots$$

so replacing z by $e^{j\omega T}$ gives the frequency-response function expressed by Equation 5.34:

$$H(e^{j\omega T}) = h[0] + h[1]e^{-j\omega T} + h[2]e^{-j2\omega T} + h[3]e^{-j3\omega T} + \dots$$

Now, if we express $H(e^{j\omega T})$ in terms of its magnitude and phase.

$$H(e^{j\omega T}) = |H(e^{j\omega T})|e^{j\Phi} \qquad (5.35)$$

we can return to the case illustrated in Figure 5.10 and, using the principle of superposition, work out the steady-state response $y[n]$ of the system to the input sinusoidal sequence $x[n] = A\cos n\omega T$. We know that

$$x[n] = A\cos n\omega T = \frac{A}{2}e^{jn\omega T} + \frac{A}{2}e^{-jn\omega T}. \qquad (5.36)$$

Using Equations 5.33 and 5.35 we can write down the response of the system to each complex-exponential component separately. The response to $\frac{A}{2}e^{jn\omega T}$ is

$$\frac{A}{2}e^{jn\omega T}H(e^{j\omega T}) = \frac{A}{2}|H(e^{j\omega T})|e^{j[n\omega T+\Phi]}. \qquad (5.37)$$

Similarly, the response to $\frac{A}{2}e^{-jn\omega T}$ is

$$\frac{A}{2}e^{-jn\omega T}H(e^{-j\omega T}) = \frac{A}{2}|H(e^{j\omega T})|e^{-j[n\omega T+\Phi]}. \qquad (5.38)$$

The response to the complex-conjugate component $\frac{A}{2}e^{-jn\omega T}$ is the complex-conjugate of the response to $\frac{A}{2}e^{jn\omega T}$. The complex-conjugate of $|H(e^{j\omega T})|e^{j\Phi}$ is $|H(e^{j\omega T})|e^{-j\Phi}$

By superposition the response to the input sinusoidal sequence $x[n] = A\cos n\omega T$ is the sum of the responses to the individual complex exponential components:

$$y[n] = \frac{A}{2}|H(e^{j\omega T})|\{e^{j[n\omega T+\Phi]} + e^{-j[n\omega T+\Phi]}\}$$
$$= A|H(e^{j\omega T})|\cos(n\omega T + \Phi). \quad (5.39)$$

Remember that we are dealing with *steady-state* behaviour. We assume that any transient behaviour caused by suddenly applying the input has died away and the system response has settled down to a steady sinusoidal output sequence of the same form as the input.

In summary, then, we can find the frequency-response function $H(e^{j\omega T})$ of any discrete-time processor by replacing z by $e^{j\omega T}$ in the transfer function $H(z)$. As in the continuous case the steady-state response of a linear discrete-time system to a sinusoidal input sequence is itself a sinusoidal sequence. The magnitude $|H(e^{j\omega T})|$ of the system frequency-response function determines the amplitude ratio of the input and output sequences, while the phase Φ determines the phase-shift of the output relative to the input. Notice that although we are dealing with discrete-time systems, the frequency response function $H(e^{j\omega T})$ is a *continuous* function of the frequency variable ω.

Worked Example 5.8

The transfer function of a two-term moving averager is

$$H(z) = \frac{1 + z^{-1}}{2}.$$

Find expressions for the magnitude and phase of the frequency response function of the averager.

Solution: Replacing z by $e^{j\omega T}$ gives the frequency-response function

$$H(e^{j\omega T}) = \frac{(1 + e^{-j\omega T})}{2}.$$

Now if we extract a factor $e^{-j\omega T/2}$ from each term on the right-hand side of this expression, we obtain

$$H(e^{j\omega T}) = \frac{(e^{j\omega T/2} + e^{-j\omega T/2})}{2} e^{-j\omega T/2}$$
$$= \cos\left(\frac{\omega T}{2}\right) e^{-j\omega T/2}$$

which can be written in magnitude and phase form,

$$H(e^{j\omega T}) = |H(e^{j\omega T})|e^{j\Phi}$$

where

$$|H(e^{j\omega T})| = \left|\cos\left(\frac{\omega T}{2}\right)\right|$$

and

$$\Phi = -\frac{\omega T}{2}$$

over the frequency range $-\pi/T < \omega < \pi/T$.

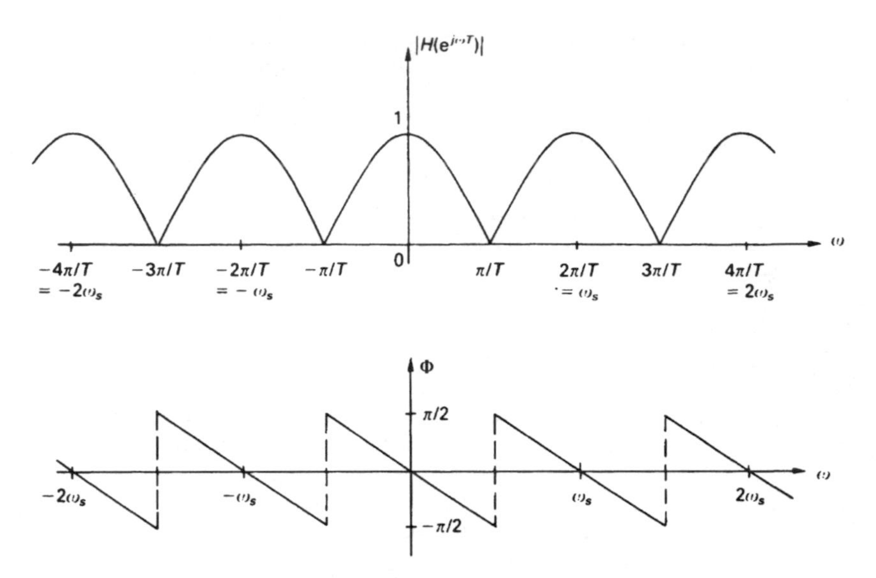

Fig. 5.11 Periodic frequency response of the two-term moving averager.

Figure 5.11 shows the magnitude and phase of the frequency response function of the moving averager. Within the frequency range $-\pi/T < \omega < \pi/T$ we see that the response has a magnitude of 1 at low frequencies and falls away to zero at $\omega = \pm\pi/T$. The linear phase characteristic shows that any signal component in the range $-\pi/T < \omega < \pi/T$ will undergo a phase shift proportional to its frequency. Remember that the frequency π/T is in fact equal to $\omega_s/2$, where $\omega_s = 2\pi f_s$ and $f_s = 1/T$ is the sampling frequency in Hz.

The sampling theorem was discussed in Chapter 1. A signal can be unambiguously represented by its samples provided the sampling frequency is greater than twice the highest frequency component present in the signal.

The frequency-response characteristics of the moving averager show that any input components with frequencies much less than the sampling frequency will appear at the output relatively unchanged in magnitude and phase. Signal components with frequencies greater than about $\omega_s/4$ on the other hand will be reduced in amplitude and shifted in phase. Now, for a signal sampled in accordance with the sampling theorem, all the significant frequency components of the original signal $x(t)$ must lie within the range $|\omega| < \omega_s/2$. For a properly sampled signal, therefore, the system will act as a *low-pass filter*.

Outside this range the response of the filter is periodic, repeating at intervals of $\omega_s = 2\pi/T$ along the frequency axis. Owing to the periodic change of sign of the function $\cos \omega T/2$ the system phase shift changes abruptly by 180° (π radians) every time the magnitude falls to zero, resulting in the periodic phase response shown in the lower part of the figure.

To see why the response should be periodic we return to the topic of aliasing introduced in Chapter 1. We saw there that sinusoidal components at frequencies differing by multiples of the sampling frequency have the same sample values. The system, however, has no knowledge of the real frequency of an input and is unable to distinguish between sinusoidal or complex components at the frequencies ω_0, $\omega_0 + \omega_s$, $\omega_0 + 2\omega_s, \ldots, \omega_0 + n\omega_s, \ldots$ The response of the system will be identical at each of these frequencies and the values of the frequency response will repeat indefinitely at intervals ω_s along the frequency axis. The resulting characteristic is thus a periodic function of the frequency variable ω.

Worked Example 5.12

A sinusoidal signal $x(t) = 3\sin 4t$ is sampled at intervals of 0.5 s to produce the sample sequence $x[n]$. This sequence forms the input to a processor defined by the transfer function

$$H(z) = \frac{z}{z - 0.5}.$$

Find the steady-state response of the processor to $x[n]$. Show that the response is identical if the input sequence is derived by sampling the signal $x(t) = 3\sin(4 + \omega_s)t$.

Solution: Replacing z by $e^{j\omega T}$ in the transfer function gives the frequency-response function

$$H(e^{j\omega T}) = \frac{e^{j\omega T}}{e^{j\omega T} - 0.5} = \frac{1}{1 - 0.5e^{-j\omega T}}.$$

Now $e^{-j\omega T} = \cos\omega T - j\sin\omega T$, so we can write

$$H(e^{j\omega T}) = \frac{1}{(1 - 0.5\cos\omega T) + j0.5\sin\omega T}$$

from which the magnitude and phase of $H(e^{j\omega T})$ are

$$|H(e^{j\omega T})| = \sqrt{\frac{1}{(1 - 0.5\cos\omega T)^2 + (0.5\sin\omega T)^2}}$$

and

$$\Phi = -\tan^{-1}\left(\frac{0.5\sin\omega T}{1 - 0.5\cos\omega T}\right).$$

We can evaluate these expressions by noting that the frequency ω of the input sinusoid is 4 rad s^{-1} and the sampling interval $T = 0.5$ s. Hence

$$\omega T = 4 \times 0.5 = 2\,\text{rad},$$
$$\sin\omega T = \sin 2 = 0.9093$$

and

$$\cos\omega T = \cos 2 = -0.4161.$$

Substituting these values into the expressions for magnitude and phase gives the amplitude ratio and the phase shift introduced by the processor:

$$|H(e^{j\omega T})| = \sqrt{\frac{1}{(1 + 0.2081)^2 + (0.4546)^2}}$$
$$= 0.77$$

and

$$\Phi = -\tan^{-1}\left(\frac{0.4546}{1 + 0.2081}\right) = -\tan^{-1}0.3763$$
$$= -0.36\,\text{rad}.$$

Now the input sample sequence is

$$x(nT) = x[n] = 3\sin n\omega T = 3\sin 2n$$

so the steady-state output is defined by the sinusoidal sequence

$$y[n] = 3 \times |H(e^{j\omega T})|\sin(n\omega T - 0.36)$$
$$= 2.3\sin(2n - 0.36).$$

If the signal $x(t) = 3\sin(4 + \omega_s)t$ is sampled at intervals of T seconds the resulting sequence will be

$$x[n] = 3\sin(4 + \omega_s)nT.$$

Now the sampling frequency $\omega_s = 2\pi/T$, so the sequence can be written

$$x[n] = 3\sin(4 + 2\pi/T)nT = 3\sin(4nT + 2\pi n).$$

But $\sin(4nT + 2\pi n) = \sin 4nT$ for $n = 0, \pm 1, \pm 2, \ldots$, so for $T = 0.5$ the sample values of the input sequence are given by

$$x[n] = 3\sin 2n$$

which is identical to the sample sequence obtained from the sinusoid $x(t) = 3\sin 4t$. Hence the response of the system to samples of $3\sin 4t$ and $3\sin(4 + \omega_s)t$ will be identical.

Exercise 5.8 Work out expressions for the magnitude response and phase shift of the system defined by the transfer function

$$H(z) = \frac{z^2 + 1}{z^2}.$$

Sketch the magnitude response function and hence show that the system behaves as a 'notch' filter for $|\omega| < \omega_s/2$ and suppresses all input components with frequencies in the region of $\omega_s/4$.

Frequency-domain interpretation of the z-plane

We have just seen that when z takes on the values $e^{j\omega T}$ we can calculate the frequency response of a discrete-time system. For different values of ω (assuming the sampling period $T = 1/f_s = 2\pi/\omega_s$ is fixed) we can conveniently express the corresponding values of $z = e^{j\omega T}$ as complex numbers of the general form $z = |z|e^{j\theta}$, where $|z| = 1$ and $\theta = \omega T$. This means that if we plot values of $z = e^{j\omega T}$ on the z-plane, as shown in Figure 5.12, each complex number will lie on the unit circle.

Recall that the unit circle plays the same role in the z-plane as the ω-axis does in the s-plane.

Let us take a few examples. If $\omega = 0$ the corresponding value of z is e^{j0},

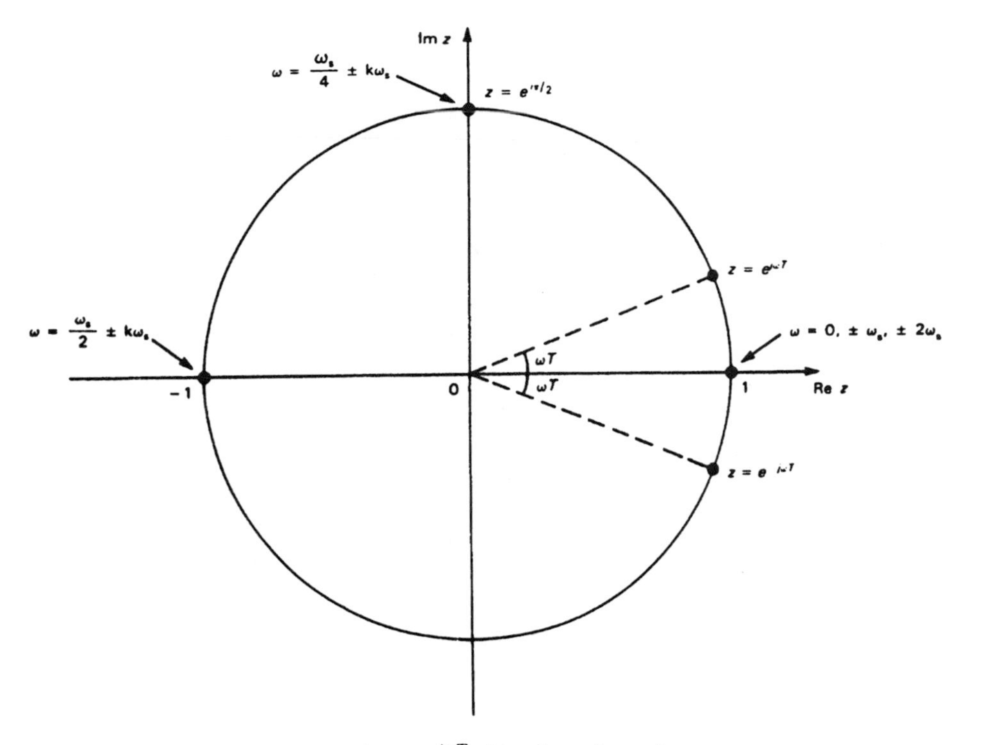

Fig. 5.12 Values of $z = e^{j\omega T}$ plotted on the z-plane.

or 1. Increasing the value of ω takes us anti-clockwise around the unit circle. At a frequency ω equal to a quarter of the sampling frequency $\omega_s/4 = \pi/2T$ we reach the value $z = e^{j\pi/2} = j$, where the imaginary part of z is 1 and the real part is zero. Increasing the frequency further to half the sampling frequency, where $\omega = \omega_s/2 = \pi/T$, takes us on to $z = e^{j\pi} = -1$. If we continue beyond this point, selecting values of $z = e^{j\omega T}$ for $\omega_s/2 < \omega < \omega_s$, we eventually complete one full rotation of the unit circle and return to our starting point $z = 1$ when we reach $\omega = \omega_s$.

Now if we increase ω yet further, beyond the sampling frequency $\omega = \omega_s$, we will start on a second trip around the unit circle. The reason is of course that the complex number $z = e^{j\omega T}$ is *periodic* in frequency, and its values repeat every time ω is increased or decreased by an amount $\omega_s = 2\pi/T$. In other words we get the same complex number z_0 for $e^{j\omega_0 T}$, $e^{j(\omega_0 \pm \omega_s)T}$, $e^{j(\omega_0 \pm 2\omega_s)T}$ and on. The point $z = 1$ on the unit circle corresponds therefore to the infinite set of frequencies $\omega = 0, \pm\omega_s, \pm 2\omega_s, \ldots \pm n\omega_s, \ldots$

We have $\omega_s = 2\pi/T$ and $e^{j2k\pi} = 1$. Therefore

$$(\omega_0 T + k\omega_s T) = (\omega_0 T + 2k\pi)$$

and

$$e^{j(\omega_0 + k\omega_s)T} = e^{j\omega_0 T} e^{j2k\pi} = e^{j\omega_0 T}.$$

We can regard a z-plane pole-zero diagram as a simplified map from which we can gain information about the transfer function at different values of z; the poles and zeros represent only the most obvious features showing where the function takes values of zero or infinity. As with the s-plane we can interpret a z-plane pole-zero pattern to give us information about the frequency response of a discrete-time system.

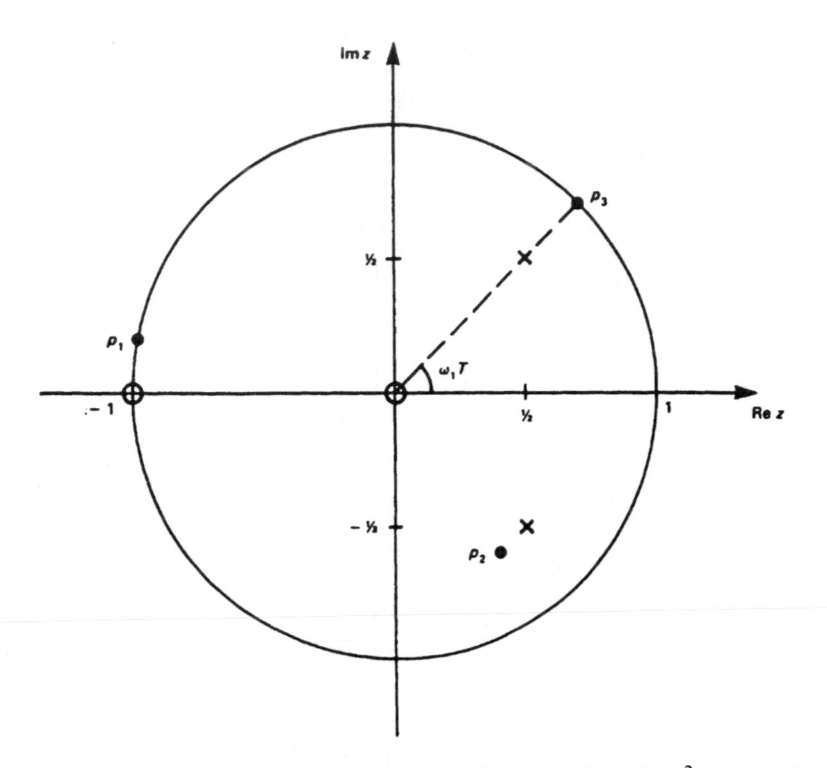

Fig. 5.13 Pole-zero diagram of the transfer function $z\,(z + 1)/(z^2 - z + 0.5)$.

To illustrate this consider the pole-zero diagram in Figure 5.13, which corresponds to the transfer function $H_3(z)$ in Worked Example 5.6, where

$$H_3(z) = \frac{z(z + 1)}{z^2 - z + 0.5}.$$

To learn about the frequency response of the system we investigate how the value of the transfer function $H_3(z)$ varies for values of z on the unit circle. If we choose a value of z that lies close to a zero, such as the value $z = p_1$ in Figure 5.13, then we would expect the corresponding value of $H_3(z)$ to be relatively small and to approach zero as z takes values closer to -1. Similarly we would expect the value of $H_3(z)$ to be relatively large around $z = p_3$ because this region of the z-plane is in the vicinity of a pole. In view of our remarks about the periodicity of the frequency response we can see that the frequency response of the system will take on relatively large values at the infinite set of frequencies $\omega_3 \pm k\omega_s$ defined by the point p_3 on the unit circle. The response will then fall away to zero at the frequencies $\omega = \omega_s/2 \pm k\omega_s$ because of the zero at $z = -1$.

This ties in with our continuous time-delay models. Recall that in a continuous system a time delay T has the Laplace transform e^{-sT}. Putting $s = j\omega$ gives the corresponding frequency domain function $e^{-j\omega T}$. Similarly a time advance would be represented by e^{sT}, or $e^{j\omega T}$ in the frequency domain.

The single zero at the origin at the z-plane in Figure 5.13 comes from the factor z in the numerator of $H_3(z)$. In frequency-response terms this corresponds to a factor $e^{j\omega T}$ in the frequency transfer function. This factor has a magnitude of 1 at all frequencies and a phase ωT which varies linearly with frequency. Such a factor corresponds to a time shift of T seconds, which is consistent with the idea that z can be regarded as a shift operator. The zero at the origin merely serves to advance the output of the system by one sampling period.

If there had been a pole at the origin this would have indicated that the output is *delayed* by one sampling period. In fact, many systems are characterized by one or more poles or zeros at the origin associated with factors z^{-m} and z^{m} in the transfer function. Their inclusion or removal changes the delay between the input and the output of a system, but their effect can be ignored for the purpose of calculating the frequency response.

Clearly no real system is able to produce an output one or more sampling periods before the input arrives. However we can remove any excess zeros at the origin (effectively by adding extra poles) with no effect on the magnitude of the frequency response. The extra phase-shift introduced is simply that due to delaying the output sequence by the appropriate number of sampling periods to ensure causal and realizable behaviour.

We can apply what we have learnt about interpreting the z-plane to the other transfer functions in Worked Example 5.6. For the two-term moving averager we have

$$H_1(z) = \frac{1 + z^{-1}}{2} = \frac{z + 1}{2z}.$$

The response at zero frequency is found by evaluating the transfer function at $z = e^{j0} = 1$, giving

$$H_1(e^{j0}) = H_1(1) = 1.$$

This single 'spot' frequency value is sufficient to establish the scale of the response magnitude, so using the principles outlined above we obtain the response shown in Figure 5.14.

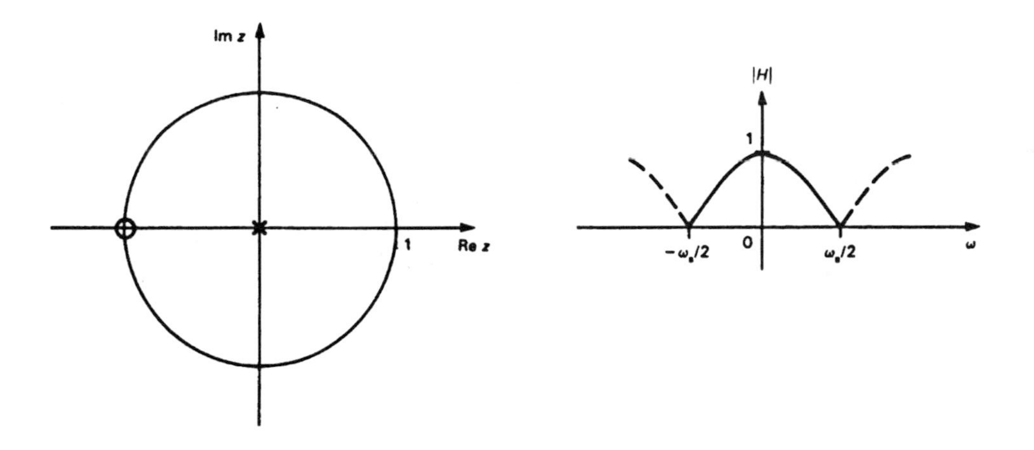

Fig. 5.14 Frequency response magnitude of two-term moving averager sketched from the z-plane pole-zero diagram.

The transfer function of the recursive system is

$$H_2(z) = \frac{z}{z - \alpha}.$$

In this case the response at zero frequency is determined by the value of α:

$$H_2(1) = \frac{1}{1 - \alpha}.$$

α also sets the pole position and we can see from Figure 5.15 how the response around zero frequency changes as α changes. For small values of α, as in Figure 5.15a, the pole lies close to the zero at the origin and its effect on the response at $z = 1$ (that is, at $\omega = 0, \pm\omega_s, \pm 2\omega_s, \ldots$ and so on) is relatively small; we can think of the effect of the pole as being largely cancelled out by the zero. Increasing α moves the pole away from the zero towards the point $z = 1$, as in Figure 5.15b, with a corresponding increase in the response magnitude at low frequencies. In Figure 5.15c the pole lies close to $z = 1$, resulting in a pronounced peak in the response around $\omega = 0$.

For $H_2(z) = z/(z - \alpha)$ the frequency response will have a maximum magnitude of $|1/(1 - \alpha)|$ for $z = 1$ (i.e. at the frequencies $\omega = 0, \pm\omega_s, \pm 2\omega_s, \ldots$) and a minimum magnitude of $|1/(1 + \alpha)|$ for $z = -1$ (at $\pm\omega_s/2, \pm 3\omega_s/2, \pm 5\omega_s/2, \ldots$)

Note that the recursive system becomes unstable if we further increase α to the extent that the pole lies beyond $z = 1$ outside the unit circle.

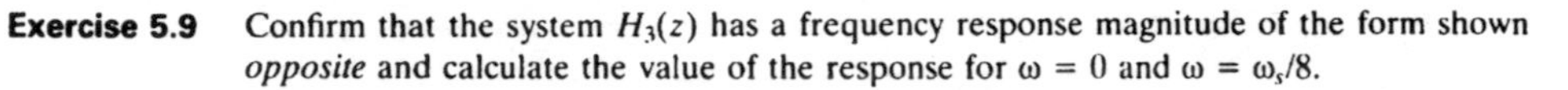

Exercise 5.9 Confirm that the system $H_3(z)$ has a frequency response magnitude of the form shown *opposite* and calculate the value of the response for $\omega = 0$ and $\omega = \omega_s/8$.

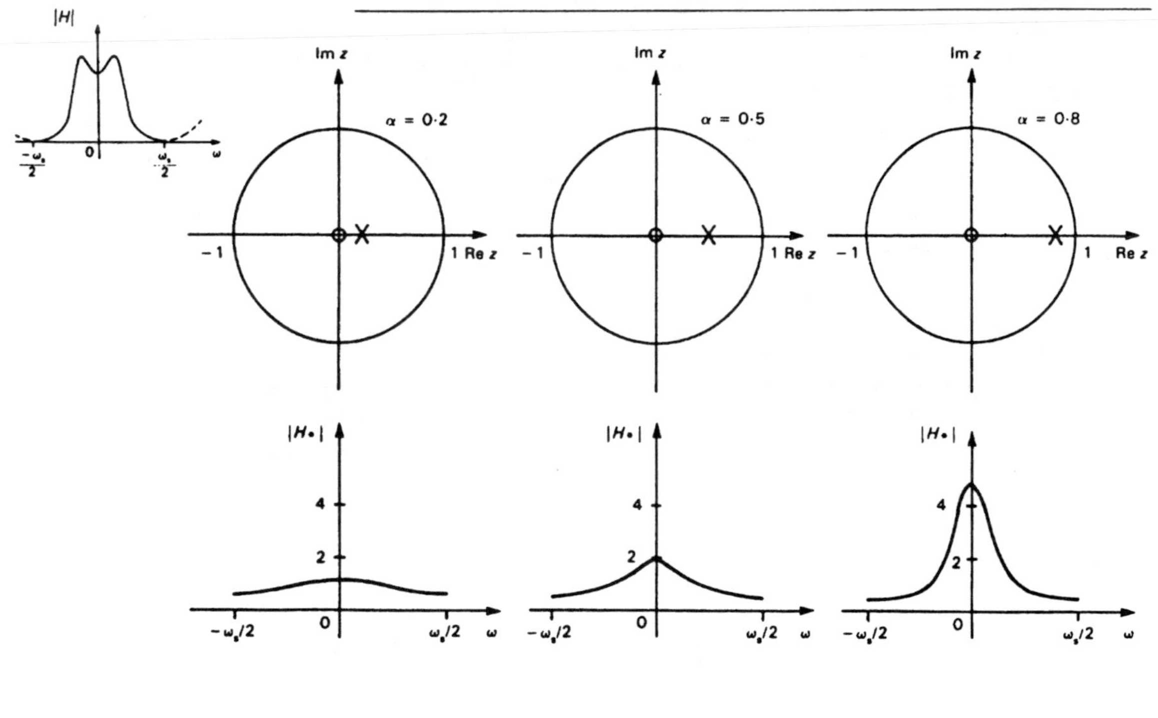

Fig. 5.15 Frequency-response magnitude associated with the transfer function $z/(z - a)$ for different values of α.

Summary

In this chapter the z-transform has been introduced as a tool to assist our understanding of discrete-time system and signal behaviour. z-transform techniques provide a way of handling sample sequences and discrete-time processors that is analogous to the Laplace transform method for continuous signals and systems.

An important result is that the convolution of two sequences turns into the multiplication of the associated z-transforms. If $X(z)$ represents the transform of an input sequence and $H(z)$ represents the transform of the unit-sample response of a processor, then the transform of the response $Y(z)$ of the processor to the input sequence is given by

$$Y(z) = H(z) \times X(z).$$

The transfer function of the processor is defined as the ratio of the transform of the output sequence to the transform of the input sequence:

$$H(z) = \frac{Y(z)}{X(z)}.$$

Hence the transfer function $H(z)$ is identical to the z-transform of the unit-sample response of the processor.

The z-transform models of many common deterministic sequences and linear processors turn out to be rational functions of the complex variable z. By identifying the values of z for which the functions become zero or infinitely large we can represent a z-plane model in terms of its poles and zeros.

In the z-plane the unit circle plays the same role as the ω-axis does in the s-plane. It represents the boundary between stability and instability. For stability all poles associated with a z-transform must lie inside the unit circle. If we evaluate a transfer function $H(z)$ on the unit-circle by replacing z by $e^{j\omega T}$, we obtain the frequency transfer function $H(e^{j\omega T})$ of the processor.

Complete information about a discrete-time system is contained in its transfer function. With practice the pole-zero diagram of $H(z)$ can be interpreted to give information about both the unit-sample response and the frequency response of the system without resorting to extensive calculations.

Problems

5.1 Derive the unit-sample response of the FIR system defined by the transfer function

$$H(z) = \frac{(z + 1)(2z - 3)(z - 2)}{z^3}.$$

5.2 An N-term moving averager has the transfer function

$$H(z) = \frac{1}{N}(1 + z^{-1} + z^{-2} + z^{-3} + \ldots + z^{-(N-1)}).$$

How many sample periods will the unit-step response of the averager take to reach its steady-state value of unity?

5.3 Find the unit-sample response of the recursive system

$$H(z) = \frac{z}{z + \alpha}$$

for the case when $\alpha = 0.25$.

5.4 Use the table of z-transform pairs in Appendix 2 to work out the z-transforms of the following sampled signals:

(a) $x(t) = 2t$, sampled at 1 Hz.
(b) $x(t) = 5e^{-10t}$, sampled at 50 Hz.
(c) $x(t) = 3(1 - e^{-5t})$, sampled at 10 Hz.

In each case the signal is assumed to be zero before $t = 0$.

5.5 Write down the first few samples of the sequences corresponding to the z-transforms:

(a) $1 + 2z^{-1} + 3z^{-3}$

(b) $\dfrac{9z}{3z - 1}$

(c) $\dfrac{z}{(z - 1)(z - 0.9)}$

(d) $\dfrac{1}{z(z^2 - 1.6z + 0.64)}$.

5.6 Use z-transforms and the method of partial fractions to find the unit-step response of the system

$$H(z) = \frac{z^2}{z^2 - 0.36}.$$

5.7 Find the poles and zeros of the systems defined by

(a) $4y[n] = x[n] + x[n - 1] - y[n - 2]$

(b) $H(z) = \dfrac{(z^3 - 2z)}{z^3 - z^2 + 2z - 2}$.

Comment on the stability of the system in each case.

5.8 Show that the following transfer function defines a processor that acts as a *band-pass* filter:

$$H(z) = \frac{z^2 - 1}{(z + 0.47 + j0.814)(z + 0.47 - j0.814)}.$$

What is the centre frequency of the filter when it is used to process signals sampled at 12 kHz?

5.9 Find the linear difference equation that describes the system $H_3(z)$ in Worked Example 5.6. Hence sketch the block diagram of the system in terms of scaling, summing and delay blocks.

5.10 Work out an expression for the magnitude of the frequency response of the moving averager defined by the transfer function

$$H(z) = \frac{1 + z^{-1} + z^{-2}}{3}.$$

If the system operates at a sampling frequency of 1 kHz calculate the value of the magnitude response at the frequencies 0, 300 Hz and 500 Hz. At what frequency in the range $0 \leq f \leq f_s$ will the magnitude response be zero?

5.11 A sinusoidal component of a signal has an amplitude of 0.5 V and a frequency of 10 Hz. The signal is sampled and applied to the input of a processor defined by the transfer function

$$H(z) = \frac{z + 1}{z - 0.9}.$$

If the sampling interval of the system is 40 ms, find the response of the processor to the input component. For what other frequencies will the response be identical?

6 Periodic Signals

Objectives

- ☐ To introduce the Fourier trigonometric series describing a periodic signal as a sum of harmonic components.
- ☐ To show that harmonic components form an orthogonal set and that the power in a periodic signal is equal to the sum of the powers in its harmonic components.
- ☐ To introduce the exponential form of the Fourier series.
- ☐ To show how the Fourier coefficients can be calculated for a given signal and presented in the form of a discrete spectrum.
- ☐ To show how a periodic signal can be described as a sum of shifted pulses and its spectrum obtained using existing results derived from the Fourier integral.
- ☐ To derive the response of a linear system to periodic inputs.
- ☐ To describe the characteristics of band-limited signals.

In this chapter we shall turn our attention to time- and frequency-domain representations of continuous-time periodic signals. It will be recalled from Chapter 1 that these are classified as *power* signals and that they are generally assumed to take values over arbitrarily long time intervals. For this reason, neither the Fourier nor the Laplace transforms are appropriate and we shall use instead the *Fourier series* description, based on a set of harmonically-related sinusoidal components. It is anticipated that many readers will be familiar with the Fourier series but be as yet unaware of the relationship between the Fourier series and the Fourier integral. We shall therefore aim to show how some periodic signals of practical interest can be analysed using transforms of basic pulse signals described in Chapter 3.

The Fourier integral can be evaluated only for *energy* signals. The Laplace Transform applies only to *causal* signals.

Strictly periodic signals

Further to the introduction in Chapter 1, we shall use the term *strictly periodic* to describe a signal with the property

$$x(t) = x(t + T_0), \text{ for all } t$$

where T_0 is the period of the signal.

The elementary waveforms shown in Figure 6.1 are strictly periodic and defined in each case by a simple mathematical expression in the range $t = -T_0/2$ to $t = T_0/2$. For example, the 'sawtooth' waveform in Figure 6.1b is defined by the ramp function

$$x(t) = 2t/T_0, \text{ for } |t| < T_0/2.$$

An interval of length T_0 that is measured from or includes the time origin $t = 0$ is known as the *fundamental interval*.

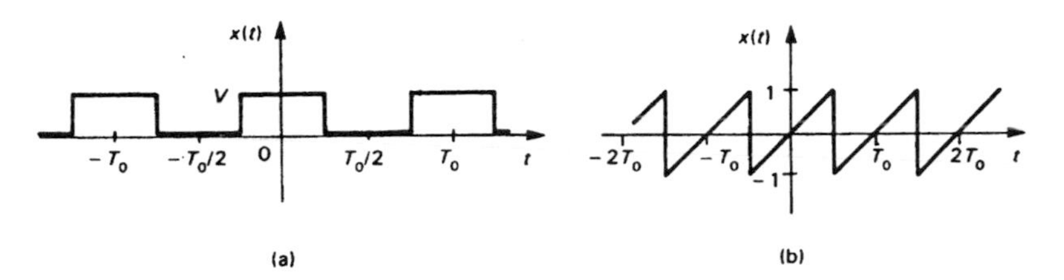

Fig. 6.1 Examples of strictly periodic signals.

Once we have defined a strictly periodic signal over the fundamental interval, or over any other interval of length equal to the period, we have sufficient information to specify its waveform for all time past, present and future.

Harmonic components

The description of periodic signals using *Fourier analysis* is based on the properties of *harmonic components*, that is components of the form $\cos m\omega_0 t$ and $\sin m\omega_0 t$. The harmonic frequencies $m\omega_0$ are integer multiples of the *fundamental frequency* ω_0. It can be shown that any *linear* combination of harmonic components is periodic with a period $T_0 = 2\pi/\omega_0$, for example, the component sum shown opposite:

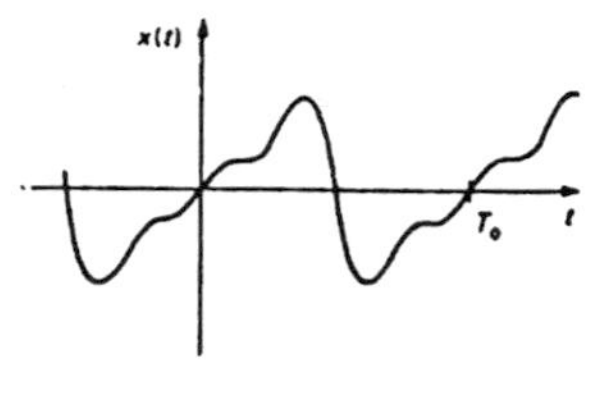

$$x(t) = \sin \omega_0 t - \tfrac{1}{2}\sin 2\omega_0 t + \tfrac{1}{3}\sin 3\omega_0 t. \tag{6.1}$$

The most general description of a signal that can be found by Fourier analysis is a summation of all possible components, cosine and sine, over all harmonic frequencies:

$$x(t) = A_0 + \sum_{m=1}^{\infty} A_m \cos m\omega_0 t + \sum_{m=1}^{\infty} B_m \sin m\omega_0 t. \tag{6.2}$$

The sum of the series is a time-dependent periodic variation with period $T_0 = 2\pi/\omega_0$.

This is the general form of the *Fourier trigonometric series*. The coefficients A_m and B_m are respectively the Fourier *cosine* and *sine coefficients* representing the amplitudes of the various components used in the description of a given signal. The constant term A_0 can be regarded as the coefficient of a cosine component at zero frequency, and thus represents the constant or *direct* component of the signal as defined in Chapter 1.

$A_m \cos m\omega_0 t = A_0$, for $m = 0$.

Although the sum of the series is a strictly periodic variation it can be used to describe the waveform of an arbitrary signal over an interval of length T_0 (see diagram). In this case however, we must remember that the description is valid *only* for $t_1 < t < (t_1 + T_0)$. Outside this interval, the values of the Fourier sum repeat periodically and so describe the *periodic extension* of the signal where the waveform repeats exactly and indefinitely at intervals of T_0 over the entire time axis.

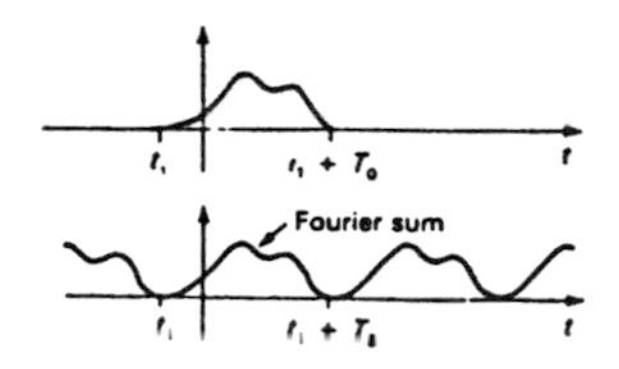

Worked Example 6.1

Identify the Fourier cosine and sine coefficients of the signal defined by Equation 6.1.

Solution: The signal is described by a Fourier *sine* series so $A_m = 0$ for all m. The fundamental component has amplitude $B_1 = 1$, the second and third harmonics are defined by $B_2 = -\frac{1}{2}$, $B_3 = \frac{1}{3}$. All other harmonics have zero value, hence $B_m = 0$ for $m \geqslant 4$.

Orthogonality and signal power

The harmonic components $\cos m\omega_0 t$ and $\sin m\omega_0 t$ are *orthogonal* over any interval of length $T_0 = 2\pi/\omega_0$. According to the definition given in Chapter 1, this means that if we take any pair of components from the following set:

We say that the components are *mutually orthogonal* and form an *orthogonal set*.

$$1, \cos\omega_0 t, \sin\omega_0 t, \cos 2\omega_0 t, \sin 2\omega_0 t, \ldots \cos m\omega_0 t, \sin m\omega_0 t, \ldots$$

their *product* will integrate or average to zero over an interval $t = t_1$ to $t = t_1 + T_0$.

We have seen that the average power of a periodic signal is given by averaging the square of the signal magnitude over a single period. If we begin with a general periodic signal defined as a Fourier sum, the square of the signal magnitude gives us a set of squared components:

$$|x(t)|^2 = A_0^2 + A_1^2\cos^2\omega_0 t + A_2^2\cos^2 2\omega_0 t + \ldots$$
$$+ B_1^2\sin^2\omega_0 t + B_2^2\sin^2 2\omega_0 t + \ldots$$

to which we must add a series of *cross-products* of the general form $2A_mB_n \cos m\omega_0 t \sin n\omega_0 t$. Now, when we average $|x(t)|^2$ over an interval of length equal to the period T_0, the cross-products will average to zero because they are orthogonal. We are then left with the sum of the powers in the individual components:

$$P_x = A_0^2 + \frac{A_1^2}{2} + \frac{A_2^2}{2} + \ldots + \frac{B_1^2}{2} + \frac{B_2^2}{2} + \ldots \tag{6.3}$$

The power of a signal described as a sum of orthogonal components is equal to the sum of the component powers.

This expression giving the power of a periodic signal in terms of its component powers is a statement of *Parseval's Theorem*. If we imagine that one or more of the components is removed from the signal, for example by the action of a selective *filter*, the signal waveform will be changed – because it now has a different composition – and its power will be reduced. However, it is not necessary to repeat the calculation over *all* the components to determine the new signal power. The orthogonal components are essentially *independent* and so the signal power will be reduced by precisely the amount resident in the missing components. As we shall see, this feature is reflected in the way that we select Fourier coefficients to represent a given signal; in any order and independently of one another.

Similarly, we can add new components and determine the increase in power without having to make reference to the existing set.

Exercise 6.1 Show that the signal defined by Equation 6.1 has an average power of 0.68 W. [According to the convention given in Chapter 1, the average power is defined for a resistive load of 1 Ω.]

The Fourier exponential series

It turns out that the analysis of a given signal into harmonic components is somewhat easier if we work with complex exponential components rather than with sines and cosines. To derive the exponential form of the Fourier series we replace each of the sines and cosines by pairs of complex exponentials:

$$A_m \cos m\omega_0 t = \frac{A_m}{2}[e^{jm\omega_0 t} + e^{-jm\omega_0 t}]$$

$$B_m \sin m\omega_0 t = \frac{B_m}{2j}[e^{jm\omega_0 t} - e^{-jm\omega_0 t}].$$

We thus obtain

$$A_m \cos m\omega_0 t + B_m \sin m\omega_0 t = X_m e^{jm\omega_0 t} + X_{-m} e^{-jm\omega_0 t}$$

where

$$X_m = \tfrac{1}{2}(A_m - jB_m);\ X_{-m} = \tfrac{1}{2}(A_m + jB_m) \tag{6.4}$$

and the direct component X_0 has the value A_0. The general form of the exponential series describing a periodic signal $x(t)$ is therefore

Notice that the summation is taken over positive and negative frequencies. It is important to remember that Equation 6.5 is a *time-domain* description, since the instantaneous values of the sum are equal to the values of the time-dependent signal $x(t)$.

$$\begin{aligned} x(t) &= X_0 + X_1 e^{j\omega_0 t} + X_2 e^{j2\omega_0 t} + \dots \\ &\quad + X_{-1} e^{-j\omega_0 t} + X_{-2} e^{-j2\omega_0 t} + \dots \\ &= \sum_{m=-\infty}^{\infty} X_m e^{jm\omega_0 t}. \end{aligned} \tag{6.5}$$

The coefficients X_m and X_{-m} are, in general, complex and specify the amplitudes and phases of the individual harmonic componments describing $x(t)$. If we return to Equation 6.4, we see that the coefficients X_m and X_{-m} must be complex conjugates. In other words, the Fourier coefficients have *conjugate symmetry*, $X_m = X^*_{-m}$, just as the values of a continuous spectrum have conjugate symmetry.

Conjugate symmetry is essential if the sum of the Fourier exponential series is to be *real* at all times, describing the variation of a real periodic signal.

Fourier analysis

We shall begin with an arbitrary signal $x(t)$ which we shall suppose has been completely defined in the range $t_1 < t < (t_1 + T_0)$. The problem is to describe the signal by a Fourier series of the general form:

$$\begin{aligned} x(t) &= X_0 + X_1 e^{j\omega_0 t} + X_2 e^{j2\omega_0 t} + \dots \\ &\quad + X_{-1} e^{-j\omega_0 t} + X_{-2} e^{-j2\omega_0 t} + \dots \end{aligned} \tag{6.6}$$

where the fundamental frequency $\omega_0 = 2\pi/T_0$. We note first of all that each of the time-varying terms on the right-hand side of Equation 6.6 can be expressed as the sum of harmonic cosine and sine components:

$$X_m e^{jm\omega_0 t} = X_m \cos m\omega_0 t + jX_m \sin m\omega_0 t$$

Therefore, if we average both sides of Equation 6.6 over a single period of $x(t)$, all of the time-varying components will average to zero and only the constant component X_0 will remain. Choosing the interval $t = t_1$ to $t = t_1 + T_0$, we obtain:

$\cos m\omega_0 t$ and $\sin m\omega_0 t$ will average to zero over any interval of length $T_0 = 2\pi/\omega_0$.

$$\frac{1}{T_0}\int_{t_1}^{t_1+T_0} x(t)\,\mathrm{d}t = X_0 \tag{6.7}$$

Notice that if we had *integrated* $x(t)$ rather than taken an average, the time-dependent terms would have disappeared as required but we would have been left with the value $X_0 T_0$.

X_0 is thus identified with the direct component of the signal, in agreement with the definition given in Chapter 1.

To find 'how much' of the component $e^{j\omega_0 t}$ is 'contained' in $x(t)$ we multiply both sides of Equation 6.6 by the complex conjugate $e^{-j\omega_0 t}$:

$$x(t)\,e^{-j\omega_0 t} = X_0 e^{-j\omega_0 t} + X_1 + X_2 e^{j\omega_0 t} + \ldots$$
$$+ X_{-1} e^{-2j\omega_0 t} + X_{-2} e^{-j3\omega_0 t} + \ldots$$

The effect of the multiplication is to cancel the time-dependence of the term involving the coefficient X_1. Hence, if we average both sides once again over a full period all the time-dependent terms disappear as before, leaving:

$$\frac{1}{T_0}\int_{t_1}^{t_1+T_0} x(t)\,e^{-j\omega_0 t}\,\mathrm{d}t = X_1. \tag{6.8}$$

Notice that Equation 6.9 need be evaluated only for positive values of m. The coefficients X_{-m} can then be found from the property of conjugate symmetry; $X_m = X^*_{-m}$.

This procedure can be repeated systematically to find each of the coefficients in turn. We can express this in a general way by writing

$$\frac{1}{T_0}\int_{t_1}^{t_1+T_0} x(t)\,e^{-jm\omega_0 t}\,\mathrm{d}t = X_m. \tag{6.9}$$

The exponential components are mutually orthogonal and form an orthogonal set like the harmonic cosines and sines.

The process of multiplying by $e^{-jm\omega_0 t}$ and averaging acts rather like a sieve from which the coefficients X_m fall out one by one. As we have already remarked, the coefficients can be calculated in any order and independently of one another. Also, the general relationships regarding signal symmetry that we noted in Chapter 3 apply here, in the same way that they did to the Fourier integral. Thus, Equation 6.9 will generate a purely *real* set of coefficients if the signal $x(t)$ has *even* symmetry about the time origin $t = 0$, and an *imaginary* set if the signal is *odd*. We can use Equations 6.4 to show that:

$$A_m = 2X_m;\ B_m = 0 \quad \text{if } X_m \text{ is } \textit{real}$$
$$A_m = 0;\quad B_m = 2jX_m \text{ if } X_m \text{ is } \textit{imaginary} \tag{6.10}$$

This satisfies the intuitive notion that an even signal will be described by cosine (even) components and an odd signal by sine (odd) components.

Worked Example 6.2

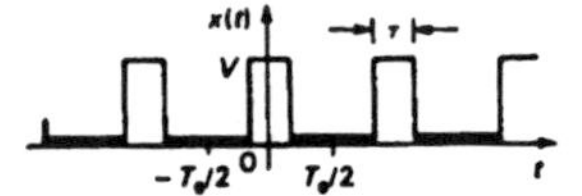

Find the Fourier coefficients describing the periodic train of rectangular pulses shown opposite.

Solution: We have

$$x(t) = V, \quad \text{for } |t| < \tau/2$$
$$= 0, \quad \text{for } \tau/2 < |t| < T_0/2.$$

Therefore the Fourier coefficients can be found by evaluating Equation 6.9 with $t_1 = -T_0/2$. We obtain

$$X_m = \frac{1}{T_0}\int_{-T_0/2}^{T_0/2} Ve^{-jm\omega_0 t}\,\mathrm{d}t = \frac{1}{T_0}\int_{-\tau/2}^{\tau/2} Ve^{-jm\omega_0 t}\,\mathrm{d}t. \tag{6.11}$$

There is an evident similarity between this expression and the Fourier integral evaluated in Chapter 3 giving the continuous spectrum of a rectangular pulse:

$$X(j\omega) = \int_{-\tau/2}^{\tau/2} Ve^{-j\omega t}\, dt. \tag{6.12}$$

Notice however that the periodic signal is a *power* signal and that Equation 6.11 represents an *average value* taken over a single period rather than a Fourier integral taken over the entire time axis.

Comparing Equations 6.11 and 6.12 we see that the coefficients can be obtained by evaluating the continuous frequency function $X(j\omega)$ at the frequencies $\omega = 0, \pm\omega_0, \pm 2\omega_0, \pm 3\omega_0, \ldots$ and then dividing by T_0. We have

$$X(j\omega) = V\tau \frac{\sin \omega\tau/2}{\omega\tau/2}.$$

Remember that the Fourier transform of a voltage pulse will be expressed in units of spectral density (V s or V/Hz). The Fourier coefficients of a periodic voltage signal have units of volts.

Hence

$$X_m = \frac{X(jm\omega_0)}{T_0} = \frac{V\tau}{T_0} \frac{\sin m\omega_0\tau/2}{m\omega_0\tau/2}.$$

Because the rectangular pulse train was defined to be an *even* function of time, its Fourier coefficients are purely real and their values can be shown on a single graph. All we have to do is to sketch the rectangular pulse spectrum $X(j\omega)$ scaled by a factor $1/T_0$ and then strike off a series of vertical lines under the $\sin x/x$ 'envelope' at a spacing $\omega_0 = 2\pi/T_0$. The result is the spectrum of the periodic signal. This is shown in Figure 6.2 for three values of the pulse repetition interval; $T_0 = 2\tau$, 5τ and 10τ.

It is assumed here that the pulse width τ remains constant while T_0 is increased, giving $\tau/T_0 = 0.5$, 0.2 and 0.1.

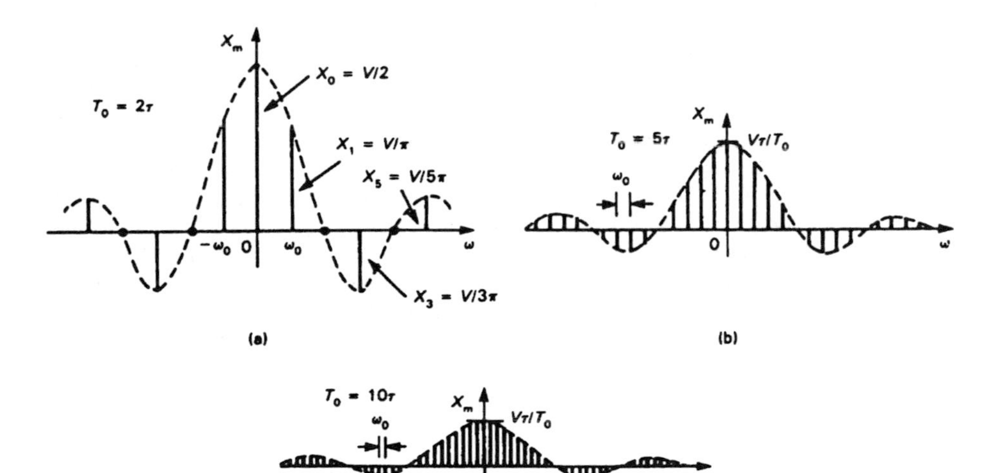

Fig. 6.2 Spectrum of a periodic rectangular pulse signal. (a) $T_0 = 2\tau$; (b) $T_0 = 5\tau$; (c) $T_0 = 10\tau$.

As T_0 takes larger values we observe two effects: the line spacing, given by $\omega_0 = 2\pi/T_0$, decreases and the magnitude of the spectrum is reduced. As the period increases therefore, more and more components become crowded into a given frequency range. Notice however that the *shape* of the spectrum does not change because this depends *only* on the rectangular pulse shape and not on the pulse repetition interval T_0.

Exercise 6.2 The squarewave signal illustrated earlier in Figure 6.1 is a special case of the rectangular pulse train with $T_0 = 2\tau$. Confirm the Fourier coefficients have the values shown in Figure 6.2a and hence show that the signal can be modelled in terms of a Fourier *cosine* series containing *odd* harmonics:

$$x(t) = \frac{V}{2} + \frac{2V}{\pi}\left[\cos\omega_0 t - \frac{1}{3}\cos 3\omega_0 t + \frac{1}{5}\cos 5\omega_0 t - \frac{1}{7}\cos 7\omega_0 t \ldots\right]$$

Exercise 6.3 Show that if the rectangular pulse train is delayed by T seconds then each of its Fourier coefficients will be multiplied by a phase factor $e^{-jm\omega_0 T}$

Fourier series and the Fourier integral

The spectrum of a periodic signal exists only at the harmonic frequencies $m\omega_0$ and for this reason it is known as a *discrete* spectrum or *line* spectrum. In the previous example we found that the *shape* of the spectrum depends only on the form of the individual pulses in the pulse train and not on the pulse spacing T_0. While this is generally true, the Fourier coefficients X_m are usually complex, so we require *two* graphs to specify the components describing a given signal; one for the magnitudes and one for the phases.

With this in mind, we shall now introduce the notation $\hat{x}(t)$ to indicate that a periodic signal has been modelled as the sum of a pulse signal $x(t)$ and its displacements of the form $x(t - kT_0)$ along the time axis as shown in Figure 6.3. The periodic signal can thus be represented by a summation of shifted pulses:

$$\hat{x}(t) = \sum_{k=-\infty}^{\infty} x(t - kT_0). \tag{6.13}$$

At this stage we have assumed that $x(t) = 0$ for $|t| > T_0/2$.

If $X(j\omega)$ is not known immediately, then it can be found from a single evaluation of the Fourier integral, an operation that is usually no more difficult than calculating the values of X_m directly using Equation 6.9.

Now, given that $x(t)$ is an *energy* signal with a continuous spectrum $X(j\omega)$, the periodic signal $\hat{x}(t)$ will be a *power* signal and described by a set of Fourier coefficients:

$$X_m = X(jm\omega_0)/T_0, \; m = 0, \pm 1, \pm 2, \ldots \tag{6.14}$$

If we express $X(j\omega)$ in polar form, $X(j\omega) = |X(j\omega)|e^{j\theta(\omega)}$, then the coefficient magnitudes and phases will be given by

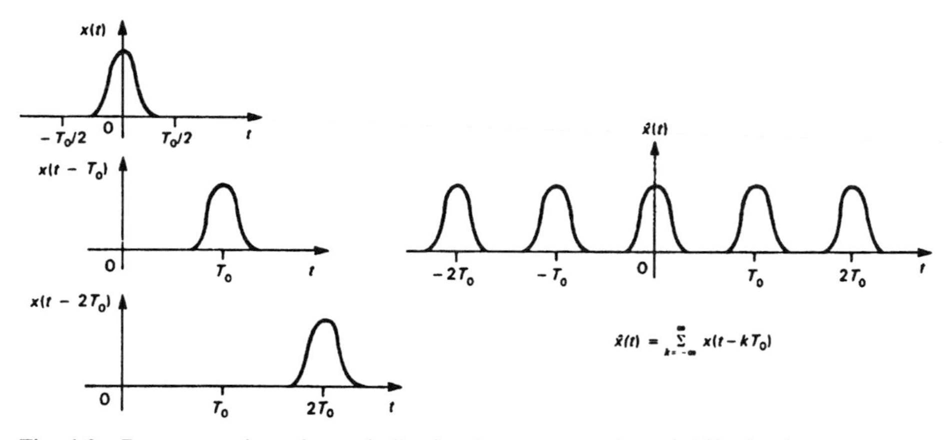

Fig. 6.3 Representation of a periodic signal as a summation of shifted pulses of the form $x(t - kT_0)$.

$$|X_m| = \frac{|X(jm\omega_0)|}{T_0};\ \theta_m = \theta(m\omega_0) \qquad (6.15)$$

In view of Equation 6.15, we can use a graphical construction to find the discrete spectrum of a signal modelled as a sum of shifted pulses. To display the magnitudes of the harmonic components we first of all sketch the magnitude $|X(j\omega)|$ of the pulse spectrum scaled by a factor $1/T_0$. Then we draw a series of vertical lines under the spectral envelope at a spacing $\omega_0 = 2\pi/T_0$. The phase spectrum is obtained in a similar fashion working from the continuous phase characteristic $\theta(\omega)$.

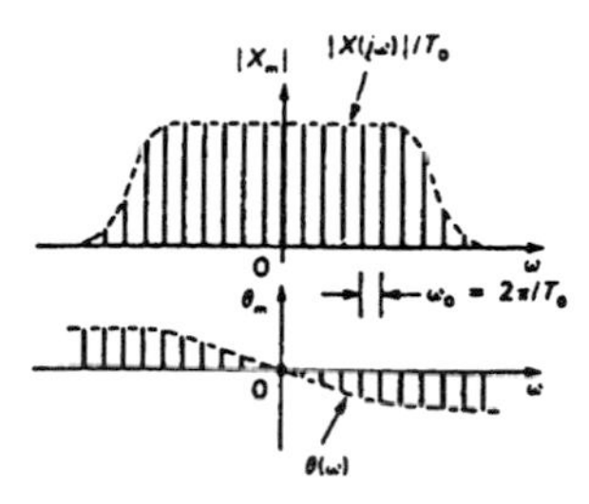

Notice that the scaling factor of $1/T_0$ applies only to the magnitude of $X(j\omega)$ and not to its phase.

Once we have established the basic form of the spectrum we can plot results for increasing values of T_0 without resorting to further calculation. As before, we observe a progressive reduction in spectral amplitude as the lines under the spectral envelope move closer together. Clearly, in the limit of very wide pulse spacing we will be left with an isolated pulse at the time origin associated with a vanishingly small line spectrum. This behaviour represents a transition from a periodic pulse train described by a Fourier series to a situation where we can model the signal as an *aperiodic* pulse defined over an interval $-T_0/2$ to $T_0/2$ where T_0 is now arbitrarily large. As we have seen, such a pulse will have finite energy and it can be fully described using the Fourier *integral*, giving a continuous spectrum with the dimensions of spectral density.

Finally, in preparation for the following section, we shall consider a periodic model of considerable practical importance. This is a train of narrow pulses modelled by the series of impulse functions shown in the margin.

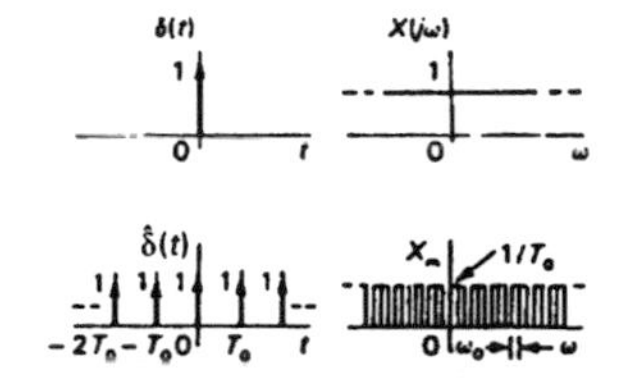

To find the Fourier series description in this case, we note the following relationship derived in Chapter 3:

$$\delta(t) \leftrightarrow X(j\omega) = 1, \text{ for all } \omega.$$

Hence, using Equation 6.15 we can write down the Fourier coefficients of the impulse train and then use Equation 6.5 to express $\hat{\delta}(t)$ as a Fourier series. We have $X_m = 1/T_0$ for all m, giving

$$\hat{\delta}(t) = \frac{1}{T_0} \sum_{m=-\infty}^{\infty} e^{jm\omega_0 t}.$$

The impulse train can thus be represented by a sum of equally weighted exponentials at the harmonic frequencies $m\omega_0$.

Input–output relationships

Many periodic signals of practical interest arise in the course of system *testing*, when the response of a system is evaluated for sinusoidal, squarewave and periodic pulse inputs. We shall suppose that a periodic signal $x(t)$ is applied to a linear system characterized by its impulse response $h(t)$ and its frequency response $H(j\omega)$. Because the system is linear, each component $X_m e^{jm\omega_0 t}$ in the input will give rise to a steady-state output component $H(jm\omega_0)\ X_m e^{jm\omega_0 t}$. We can therefore express the periodic output signal as a superposition of components:

The response of a linear system to a periodic input is periodic with the same period as the input.

$$y(t) = \sum_{m=-\infty}^{\infty} H(jm\omega_0) X_m e^{jm\omega_0 t}. \tag{6.16}$$

We can also express this result in terms of the Fourier coefficients of the output $y(t)$:

$$Y_m = H(jm\omega_0) X_m \tag{6.17}$$

showing that each Fourier coefficient of the output is obtained by multiplying the corresponding Fourier coefficient of the input by an appropriate value of $H(j\omega)$.

Worked Example 6.3 The impulse response of a system is measured by applying a train of narrow pulses to its input, modelled by the impulse train $\hat{\delta}(t)$ described in the previous section. Write down the Fourier coefficients describing the periodic output of the system.

Solution: The impulse train is characterized by a set of Fourier coefficients $X_m = 1/T_0$. For a time-invariant linear system having a frequency response $H(j\omega)$, the output will have coefficients defined by Equation 6.17. Putting $X_m = 1/T_0$ we obtain

$$Y_m = H(jm\omega_0)/T_0 \tag{6.18}$$

To visualize the periodic signal appearing at the system output we recall that an impulse applied to the input at time $t = kT_0$ produces a delayed impulse response of the form $h(t - kT_0)$. The response to a train of impulses $\hat{\delta}(t)$ will therefore be a superposition of shifted impulse responses described by the summation

Remember that we can also describe the output in terms of the *convolution*: $y(t) = \hat{\delta}(t) * h(t)$.

$$y(t) = \hat{h}(t) = \sum_{k=-\infty}^{\infty} h(t - kT_0).$$

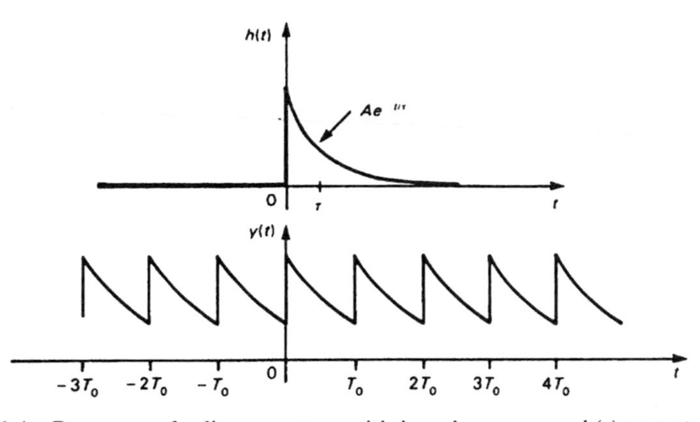

Fig. 6.4 Response of a linear system with impulse response $h(t)$ to a periodic impulse train.

Figure 6.4 shows an example of the output waveform for a causal system characterized by an impulse response

$$h(t) = Ae^{-t/\tau} \quad \text{for } t \geqslant 0. \tag{6.19}$$

It is clear that the actual form of the output will depend on the choice of T_0 compared with the time constant τ of the response. In particular, if T_0 is made too small then considerable 'overlap' will occur in the shifted responses as shown in the illustration. However, the general form of the Fourier series describing the output remains valid for all values of T_0 with Fourier coefficients given by Equation 6.18:

Notice that if overlap occurs the periodic output waveform will *not* be equal to $h(t)$ for $0 < t < T_0$. To find the waveform we must make a sketch and then add together the shifted responses.

$$Y_m = H(jm\omega_0)/T_0$$

where, in this case

$$H(j\omega) = \frac{A\tau}{(1 + j\omega\tau)}.$$

Figure 6.5 (overleaf) shows the spectrum of the periodic output obtained using the magnitude characteristic $|H(j\omega)|$ and the phase shift $\theta(\omega)$ of the system frequency response. In the first example, Figure 6.5a, the pulse spacing T_0 is equal to the time-constant τ of the response so that it is almost impossible to distinguish the individual pulses in the output waveform. In Figure 6.5b, the pulse spacing has been increased by a factor of 4. The spectral lines are thus more closely spaced and the overall spectral magnitude has been reduced while the output pulses are now more clearly separated along the time axis.

As a rule, we should allow a minimum of about five time-constants to ensure that the system responses are clearly separated in time.

Bandlimited signals

Figure 6.6 (overleaf) shows the waveform and spectrum of a typical *band-limited signal* defined over an interval $t = 0$ to $t = T_0$. Because the signal is bandlimited its spectrum extends to a maximum frequency $M\omega_0$ which defines the signal *bandwidth* in rad s^{-1}. The signal can thus be represented by a *finite* Fourier series of the general form:

A bandlimited signal defined over a finite interval can always be specified using a finite set of Fourier coefficients.

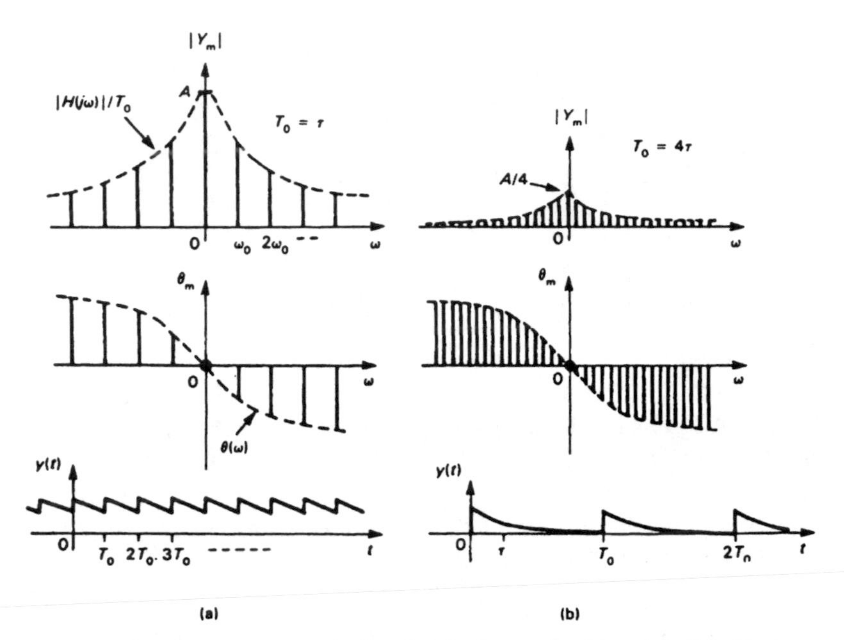

Fig. 6.5 Spectrum of a periodic signal shown in Fig. 6.4. (a) $T_0 = \tau$; (b) $T_0 = 4\tau$.

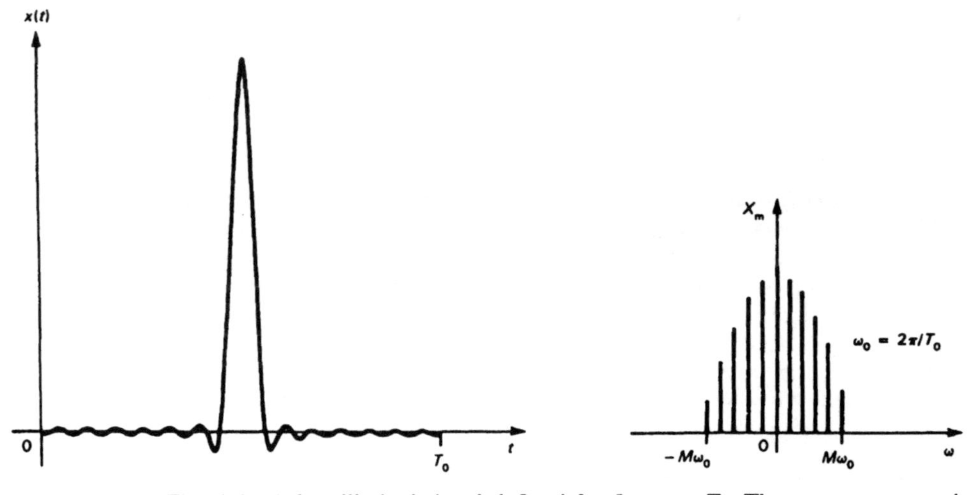

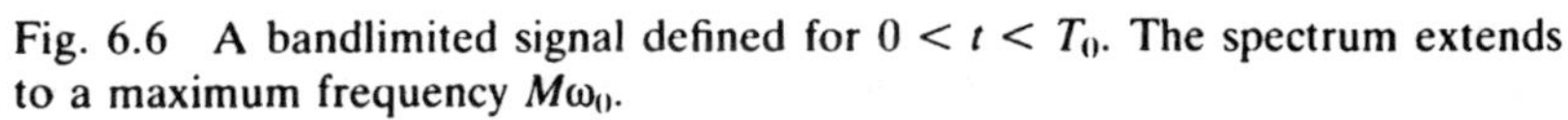

Fig. 6.6 A bandlimited signal defined for $0 < t < T_0$. The spectrum extends to a maximum frequency $M\omega_0$.

$$x(t) = \sum_{m=-M}^{M} X_m e^{jm\omega_0 t} \text{ for } 0 < t < T_0. \tag{6.20}$$

If we include only the a.c. components, we see that the Fourier series contains a total of $N = 2M$ terms. For a signal with bandwidth $2\pi f_B$, the value of M can be found from the relationship $2\pi f_B = M\omega_0$, where $\omega_0 = 2\pi/T_0$. We thus obtain $M = f_B T_0$ and reach the following important conclusion:

> A signal of bandwidth f_B defined over an interval of length T_0 can be represented by $2M$ harmonic components, where $M = f_B T_0$.

When we use the finite Fourier series to model the behaviour of a measurable signal we do not mean to imply that the signal is *strictly* bandlimited, with its components taking zero values beyond a well-defined frequency. Rather, we use the model to indicate that all the significant components of a particular signal lie within a certain frequency range and that components beyond this range are small enough to be ignored for practical purposes. The following example suggests how a bandlimited signal might be generated in practice – by transmitting a signal with arbitrarily wide bandwidth through a low-pass filter.

The squarewave signal considered in Exercise 6.2 is applied to an ideal low-pass system with a gain of unity in the pass-band and a cut-off frequency of 60 kHz. If the input has a period of 100 µs, express the filter output as a Fourier series.

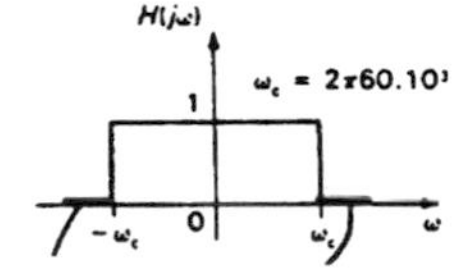

Solution: The input has a fundamental frequency $f_0 = 1/10^{-4} = 10\,\text{kHz}$. The signal is composed entirely of *odd* harmonics, so the highest harmonic in the output will be the fifth at a frequency of 50 kHz. The general form of the output Fourier series is therefore given by Equation 6.20 with $M = 5$. Using the cosine form of the series we have

$$y(t) = \frac{V}{2} + \frac{2V}{\pi}\left(\cos\omega_0 t - \frac{1}{3}\cos 3\omega_0 t + \frac{1}{5}\cos 5\omega_0 t\right).$$

The output signal in the previous example is shown, defined by the Fourier series over the range $|t| < T_0/2$. We see that the waveform has the characteristic 'ringing' or oscillatory appearance of a signal described by a small number of harmonic components. Notice in particular that the sharp transitions that marked the behaviour of the input are completely absent in the output which now has a finite 'rise-time'. The effect is, of course, due entirely to the removal of high-frequency components by the filter, resulting in an output that changes relatively slowly compared with the input.

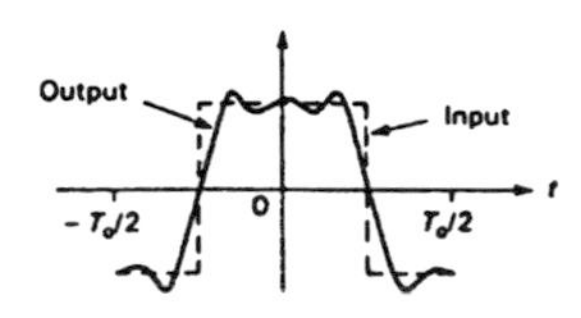

Summary

We have shown that a strictly periodic signal can be described as a sum of orthogonal harmonic components. The calculation of signal power is greatly simplified when orthogonal components are used and the components describing a given signal can be found systematically and independently of one another using the techniques of Fourier analysis. When we plot the Fourier coefficients of a signal versus frequency the result is a discrete spectrum or line spectrum, specifying the amplitudes and phases of the individual harmonic components. For a signal described as a sum of shifted pulses, the form of the spectrum depends only on the pulse shape and we have seen that in many cases the spectrum can be found from a graphical construction, using existing results obtained from the Fourier integral. Given the form of the spectrum it becomes possible to evaluate the effect of linear systems on a periodic signal and it was shown that in certain cases the output may be described as a series of overlapping pulses. Regardless of the degree of overlap however, the spectrum of the output can always be found graphically without the need for tedious calculation of individual Fourier coefficients.

It was shown finally that if a signal is band-limited with bandwidth f_B, a signal record of duration T_0 can be represented by a finite number. $M = 2f_BT_0$, of Fourier components. We shall meet this idea again in the final chapter where we shall be using periodic models in the context of discrete-time signals and systems.

Problems

6.1 Find the period T_0 and the complex coefficients X_m of the periodic signal $x(t) = 1 + 2\cos 10t + 3\sin 15t$.

6.2 Find the average power in the signal described in problem 6.1.

6.3 Show that the power in a signal represented by an exponential Fourier series is given by this form of Parseval's theorem:

$$P_x = X_0^2 + 2\sum_{m=1}^{\infty}|X_m|^2.$$

A squarewave signal exhibits half-wave symmetry.

6.4 Show that if a signal satisfies the condition for *half-wave symmetry*: $x(t) = -x(t \pm T_0/2)$, its Fourier series will contain only odd harmonic terms.

6.5 Find the coefficients of the exponential Fourier series describing the sawtooth waveform illustrated in Fig. 6.1.

6.6 Show that the conclusions of Exercise 6.2 apply to any time-shifted periodic signal.

6.7 For the case considered in Worked Example 6.4, show that just under 97% of the input power is transmitted to the output.

6.8 A periodic pulse train is applied to the input of a two-pole system defined by the transfer function

$$H(s) = \frac{\omega_c^2}{s^2 + 2\omega_c s + \omega_c^2}$$

(a) Find the Fourier coefficients of the output if the input can be modelled as a series of impulse functions with area 10^{-4} Vs.

(b) If $\omega_c = 10^3$, suggest how the pulse spacing T_0 may be chosen to avoid serious errors in the measurement of the impulse response due to pulse 'overlap' in the output.

Fourier Analysis of Discrete-time Signals and Systems 7

Objectives

- □ To introduce the discrete-time Fourier transform and explain how it may be used to find the steady-state response of a discrete-time system to a given input sequence.
- □ To show how to calculate the spectrum of a discrete-time signal or sequence.
- □ To describe how the spectrum of a discrete-time signal can be approximated by inspecting the pole-zero diagram of the signal z-transform.
- □ To describe some of the properties of the discrete-time Fourier transform.
- □ To derive the inverse discrete-time Fourier transform, whereby the values of a sequence are recovered from its spectrum.
- □ To look further at the implications of the sampling theorem and explain how the spectrum of a signal is related to the spectrum of its sample values.
- □ To describe the effects of truncating a signal to give a finite-length signal record and to show, in particular, that signal truncation imposes a limit on the resolution that can be achieved when calculating the signal spectrum.
- □ To introduce the topic of frequency-domain sampling.
- □ To introduce the discrete Fourier transform and show how it is used to relate the values of a finite-length sequence to the sample values of its spectrum.

Our first aim in this chapter is to establish the role of frequency-domain models in the study of discrete-time signals and systems. We then continue to describe how the spectrum of an experimental signal is related to the spectrum of its samples. The final part of the chapter deals with some of the practical aspects of calculating signal spectra and provides an introduction to the discrete Fourier transform or DFT.

The DFT provides the basis of much of the analysis and processing of signals that is carried out nowadays using digital computers and dedicated digital processing devices. When supplied with a suitable program, a digital computer will operate on samples of experimental data signals to perform a spectral analysis, extract information or act as a filter. Results can be plotted to the screen or output in hard-copy form. If found to be unsatisfactory, the program might be modified or extended, providing considerable flexibility compared with the alternative, analogue-based approach, which may involve redesign, construction and testing of specialist circuitry.

The vast literature devoted to the use and application of digital signal processing systems reflects the rapid development of digital computer and integrated circuit technology over the last twenty years. Nevertheless, if we remove the emphasis from the purely computational aspects, we find that the study of these systems depends ultimately on an understanding of linear system theory, the use of transform methods and an appreciation of the mathematical models that we have been concerned with in previous chapters. While developing these ideas we have found it convenient to treat continuous-time and discrete-time signals and systems more or less separately for the purposes of analysis and discussion. In the following sections we shall aim for a more unified treatment which brings out more clearly the relationship between the continuous-time and discrete-time domains.

The Laplace and z-transforms

We introduced the z-transform in Chapter 5 and described some of its properties. In common with all other transforms met so far, the z-transform enables us to replace a convolution in the time domain with a product in the transform domain. So if a signal described by a sequence $x[n]$ is applied to a system with unit-sample response $h[n]$, we can transform the convolution product

$$y[n] = h[n] * x[n] \tag{7.1}$$

to an algebraic product of transforms

$$Y(z) = H(z) \times X(z) \tag{7.2}$$

and then invert $Y(z)$ in order to find the successive values of the output sequence $y[n]$.

We noted the close correspondence between the z-transform approach and the use of the Laplace transform with continuous-time signals and systems. In the continuous-time case, the equation of transforms becomes

$$Y(s) = H(s) \times X(s). \tag{7.3}$$

A causal signal or sequence is defined to be zero for $t < 0$ or $n < 0$.

When the Laplace transform is used as a tool for solving differential equations, the steady-state response is sometimes referred to as the *particular integral* and the transient response as the *complementary function.*

It is important to remember that we defined both the Laplace and z-transforms in such a way that their application is restricted to *causal* signals. This means that when we invert the transform of the output, $Y(z)$ or $Y(s)$, we are effectively finding the response of a system to a signal switched on at $t = 0$ or $n = 0$. Now, because the transfer function provides a complete description of the transient and steady-state behaviour of the system, the response always comprises the sum of two parts. One part, known as the transient response, has a form determined by the properties of the system and eventually decays to zero if the system is stable. When sufficient time has elapsed we observe the second part, the steady-state response, which depends on the particular input $x(t)$ or $x[n]$.

The Fourier transform and the DTFT

In practice it may be a rather lengthy and difficult task to solve for the output and to separate the steady-state part of the response. However, if only the steady-state part is required, it usually easier and more direct to set up a steady-state sinusoidal analysis in which the input is modelled as a superposition of steady sinusoidal components at different frequencies. We recall the basis of this approach, which is that the steady-state response of a linear system to an input sinusoid is a sinusoid of the same frequency, but differing in amplitude and phase. When a linear sum of sinusoidal components is applied to a stable linear system, the system responds by modifying the amplitude and shifting the phase of each input component by an amount depending on its frequency. The linear superposition of all the modified components constitutes the steady-state output of the system.

When dealing with continuous-time signals and systems the input signal $x(t)$ is represented by its Fourier spectrum $X(j\omega)$ and the system by its steady-state frequency-response function $H(j\omega)$. The product of $X(j\omega)$ and the frequency-response function determines the spectrum of the output:

$$Y(j\omega) = H(j\omega) \times X(j\omega) \tag{7.4}$$

which, on inversion, gives the waveform of the steady-state output signal, $y(t)$.

It was shown in Chapter 4 that the frequency-response function $H(j\omega)$ can be obtained by evaluating the Laplace transfer function $H(s)$ along the line $s = j\omega$, marking the boundary between stable and unstable behaviour in the complex s-plane. As explained in Chapter 5, the equivalent operation in the discrete-time case corresponds to finding the values of the transfer function $H(z)$ on and around the unit circle in the complex z-plane. Thus, if we substitute $z = e^{j\omega T}$ in $H(z)$, we obtain the discrete-time frequency-response function $H(e^{j\omega T})$:

The unit circle represents the boundary of stability in the complex z-plane.

$$H(e^{j\omega T}) = H(z)|_{z=e^{j\omega T}} \tag{7.5}$$

Now, we know that the frequency-response function of a continuous-time system can be found directly by applying the Fourier transform to the unit-impulse response, $h(t)$:

$$h(t) \overset{\text{FT}}{\longleftrightarrow} H(j\omega) = \int_{-\infty}^{\infty} h(t)\,e^{-j\omega t}\,dt \tag{7.6}$$

The frequency-response function of a discrete-time system is similarly related to its unit-sample response $h[n]$ by a formal transformation. We write

$$h[n] \overset{\text{DTFT}}{\longleftrightarrow} H(e^{j\omega T}) \tag{7.7}$$

where DTFT denotes the *discrete-time Fourier transform*. We can bring out the nature of this transformation by taking a two-step approach. First, we apply the z-transform to $h[n]$ to obtain the system transfer function $H(z)$:

In some textbooks, $H(e^{j\omega T})$ is known simply as the Fourier transform of $h[n]$. By using the alternative term discrete-time Fourier transform we are reminded that we are operating on the values of a discrete-time signal.

$$h[n] \overset{z}{\longleftrightarrow} H(z) = \sum_{n=-\infty}^{\infty} h[n]\,z^{-n}. \tag{7.8}$$

Then we set $z = e^{j\omega T}$, giving

$$h[n] \overset{\text{DTFT}}{\longleftrightarrow} H(e^{j\omega T}) = \sum_{n=-\infty}^{\infty} h[n]\,e^{-jn\omega T}. \tag{7.9}$$

The series form of $H(e^{j\omega T})$ which appears in Equation 7.9 was originally given in Equation 5.34 of Chapter 5.

The frequency-response function $H(e^{j\omega T})$ characterizes the steady-state response of a discrete-time system to a complex exponential input component $w[n] = e^{jn\omega T}$. $H(e^{j\omega T})$ is in general a complex function which is conveniently expressed in polar form:

$$H(e^{j\omega T}) = |H(e^{j\omega T})|\,e^{j\Phi(\omega)}. \tag{7.10}$$

When we plot a graph showing the magnitude and phase of $H(e^{j\omega T})$, as in Figure 7.1, the horizontal axis is scaled with respect to the sampling frequency

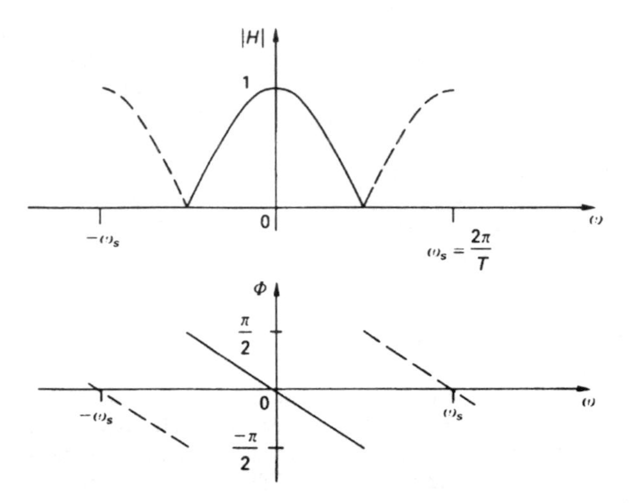

Fig. 7.1 Frequency response magnitude and phase for a discrete-time low-pass filter.

It was shown in Chapter 5 that the frequency-response function $H(e^{j\omega T})$ is a periodic function of frequency which repeats its values at intervals of ω_s along the frequency axis.

$\omega_s = 2\pi f_s$, or equivalently, the sampling interval, $T = 1/f_s$. We show here a low-pass system, but in practice the frequency response could take any number of forms, band-pass, high-pass, notch, depending on the location of the poles and zeros of the transfer function.

If we know the frequency response of a discrete-time system then we can, in principle, determine the steady-state response of the system to an arbitrary input sequence, $x[n]$, represented as a superposition of complex exponential sequences or sinusoidal component sequences of the type $\sin n\omega T$. As in the continuous-time case, the frequency analysis of a discrete-time signal or sequence involves transforming a time-domain model to an equivalent frequency-domain model. The appropriate transformation is the DTFT which, when applied to a sequence $x[n]$, gives the *spectrum of the sequence*, denoted by $X(e^{j\omega T})$:

We obtain the spectrum of a sequence $x[n]$ by substituting $z = e^{j\omega T}$ in its z-transform, $X(z) = \Sigma x[n] z^{-n}$.

$$x[n] \xleftrightarrow{\text{DTFT}} X(e^{j\omega T}) = \sum_{n=-\infty}^{\infty} x[n] e^{-jn\omega T}. \tag{7.11}$$

The spectral function $X(e^{j\omega T})$ serves as the discrete-time counterpart to the Fourier spectrum $X(j\omega)$ of a signal $x(t)$ defined by

$$x(t) \xleftrightarrow{\text{FT}} X(j\omega) = \int_{-\infty}^{\infty} x(t) e^{-j\omega t} dt. \tag{7.12}$$

Whereas $X(j\omega)$ may be found by evaluating the Laplace transform $X(s)$ along the ω-axis in the complex s-plane, $X(e^{j\omega T})$ represents the values of the z-transform $X(z)$ on the z-plane unit circle. $X(e^{j\omega T})$ is, in general, a complex function of frequency with magnitude $|X(e^{j\omega T})|$ and phase $\Phi(\omega)$:

$$X(e^{j\omega T}) = |X(e^{j\omega T})| e^{j\Phi(\omega)}. \tag{7.13}$$

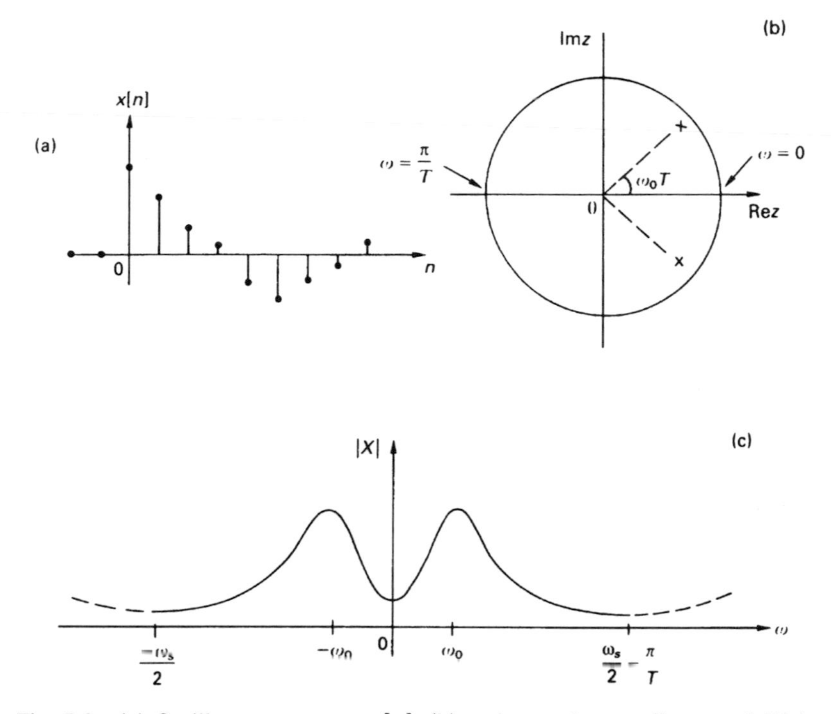

Fig. 7.2 (a) Oscillatory sequence $x[n]$; (b) z-plane pole-zero diagram of $X(z)$; (c) An approximate plot of the spectral magnitude $|X(e^{j\omega T})|$ inferred from the pole-zero diagram.

$X(e^{j\omega T})$, like the frequency-response function $H(e^{j\omega T})$, is a continuous function of frequency. Provided the pole-zero plot of the signal z-transform $X(z)$ is not too complicated, the graph of the spectral magnitude can be approximated in a manner identical to that described in Chapter 5 for the frequency-response function. For example, Figure 7.2 shows a decaying oscillatory sequence, $x[n]$, characterized by a pair of complex-conjugate poles at an angular displacement $\pm\omega_0 T$ measured from the horizontal axis in the z-plane. Because of the close proximity of the poles and the unit circle, the spectral magnitude is strongly peaked in the vicinity of the frequency ω_0. Thus, as we move around the unit circle in an anticlockwise direction, the spectral magnitude rises from a relatively low value at $\omega = 0$, passes through a maximum as we skirt the pole and then falls away again as we approach the frequency π/T, equivalent to half the sampling frequency. The sketch of the spectral magnitude can be completed for $0 < \omega < -\pi/T$ by traversing the circle in a clockwise direction, but the only way we can evaluate the spectrum outside the fundamental range $-\pi/T < \omega < \pi/T$ is to make repeated trips around the unit circle. As a result, we find that the spectrum of a sequence, like the frequency-response function $H(e^{j\omega T})$, is a periodic function of frequency which repeats its values whenever ω is increased or reduced in multiples of the sampling frequency $\omega_s = 2\pi/T$.

When we set $z = e^{j\omega T}$ in the z-transform of a signal $x[n]$, we are expressing the signal in terms of steady sinusoidal sequences. The fact that $X(e^{j\omega T})$ is a continuous function of frequency implies that the components of $x[n]$ are continuously distributed over the frequency axis. We can therefore think of $X(e^{j\omega T})$ as a *spectral density* function.

The graph of a spectrum $X(j\omega)$ can be approximated in a similar way, by taking note of the location of the poles and zeros of a signal with respect to the ω-axis in the complex s-plane.

The spectrum of the sequence in Figure 7.2 is a complex function of frequency. We show only the spectral magnitude here.

Exercise 7.1 Sketch an approximation to the spectral magnitude of a discrete-time signal represented by the pole-zero diagram shown opposite.

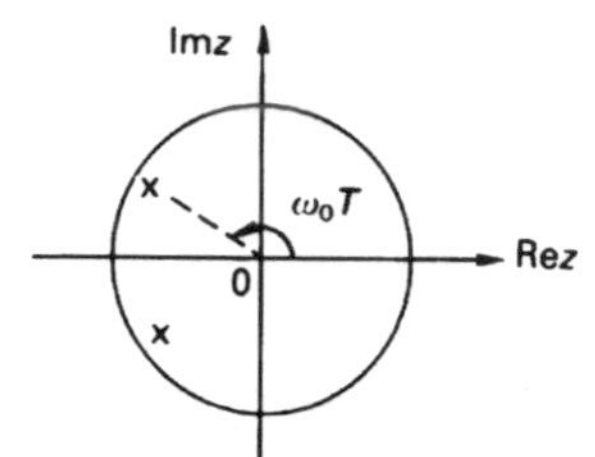

We now have an almost complete correspondence between the frequency-domain models of continuous-time and discrete-time signals and systems. To describe the filtering effect of a linear discrete-time system on the frequency components of an input sequence, we multiply the input spectrum $X(e^{j\omega T})$ with the system frequency-response function $H(e^{j\omega T})$, giving the output spectrum

$$Y(e^{j\omega T}) = H(e^{j\omega T}) \times X(e^{j\omega T}). \tag{7.14}$$

This expression is a discrete-time counterpart to Equation 7.3. The next step is to review some of the properties of the discrete-time Fourier transform. We shall then be in a position to define an appropriate *inverse* transform which would enable us to recover the values of the steady-state output sequence $y[n]$ from its spectrum $Y(e^{j\omega T})$.

Remember that when we multiply complex quantities we multiply their magnitudes and add their phases. The spectrum $Y(e^{j\omega T})$ therefore has a magnitude given by the product of $|X(e^{j\omega T})|$ and $|H(e^{j\omega T})|$. The phase of $Y(e^{j\omega T})$ is found by adding the phase characteristics of $X(e^{j\omega T})$ and $H(e^{j\omega T})$.

Further properties of the DTFT

We shall first demonstrate the application of the DTFT to the finite-length sequence shown opposite. This sequence, defined by the three non-zero values $x[0] = \frac{1}{4}$, $x[1] = \frac{1}{2}$ and $x[2] = \frac{1}{4}$, is sometimes known as the Hanning sequence.

Worked Example 7.1 Calculate the spectrum $X(e^{j\omega T})$ of the Hanning sequence and sketch the spectral magnitude and phase for $|\omega| < \omega_s/2$. Indicate the form of the periodic extension of $X(e^{j\omega T})$ for $|\omega| > \omega_s/2$.

Solution: The spectrum is given by the finite summation:

$$\begin{aligned} X(e^{j\omega T}) &= \sum_{n=0}^{2} x[n]\,e^{-jn\omega T} \\ &= \tfrac{1}{4} + \tfrac{1}{2}e^{-j\omega T} + \tfrac{1}{4}e^{-j2\omega T} \\ &= \tfrac{1}{2}(1 + \tfrac{1}{2}e^{j\omega T} + \tfrac{1}{2}e^{-j\omega T})e^{-j\omega T} \\ &= \tfrac{1}{2}(1 + \cos\omega T)e^{-j\omega T}. \end{aligned}$$

The spectral magnitude and phase may be found by comparing this expression with the polar form of $X(e^{j\omega T})$:

$$X(e^{j\omega T}) = |X(e^{j\omega T})|e^{j\Phi(\omega)}$$

whence

$$|X(e^{j\omega T})| = \tfrac{1}{2}(1 + \cos\omega T)$$

and

$$\Phi(\omega) = -\omega T.$$

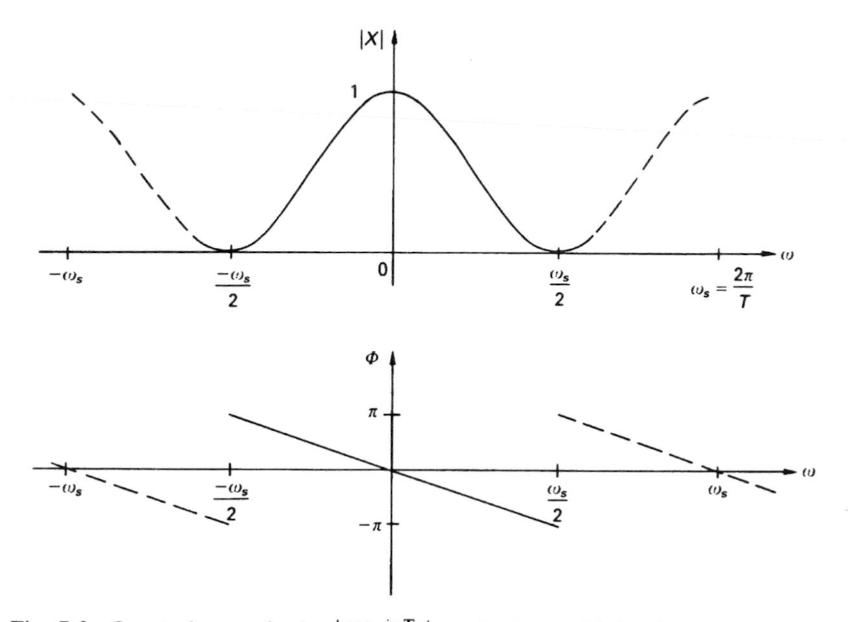

Fig 7.3 Spectral magnitude $|X(e^{j\omega T})|$ and phase $\Phi(\omega)$ of the Hanning sequence.

The graph of the spectrum is shown in Figure 7.3. The phase varies linearly as $-\omega T$ over all frequencies. But, because we always express the phase in terms of its principal value in the range $[-\pi, \pi]$, the phase characteristic behaves in periodic 'sawtooth' fashion when we plot the spectrum outside the fundamental interval $\omega = -\omega_s/2$ to $\omega = \omega_s/2$.

The spectrum of the Hanning sequence is sometimes known as the raised-cosine spectrum. The Hanning filter has a frequency-response of this form and is widely used as a low-pass filter for smoothing the values of an input sequence.

Convergence of the DTFT

It was noted in Chapter 3 that the continuous-time Fourier integral always converges to a well-defined function of frequency when applied to finite-energy signals. Similarly, we can ensure that the sum of exponential terms defining the DTFT of a sequence will be finite for all frequencies by restricting its application to aperiodic sequences which decay to zero for increasing values of n. Now, because we approached the DTFT from the point of view of the z-transform, we have so far assumed that all sequences have been causal sequences. But, like the continuous-time Fourier transform, the DTFT can be more widely used, subject to the following condition which is *sufficient* to ensure convergence:

$$\sum_{n=-\infty}^{\infty} |x[n]| < \infty. \tag{7.15}$$

Compare this expression with Equation 3.20 in Chapter 3.

A sequence with this property is said to be *absolutely summable*. It can be shown further that an absolutely summable sequence is necessarily aperiodic and possesses finite energy, that is

$$\sum_{n=-\infty}^{\infty} |x[n]|^2 < \infty. \quad (7.16)$$

It is clear than any finite-length sequence will be absolutely summable in the sense of equation 7.15 and thus possess a DTFT. Otherwise, for infinite sequences, the conditions for convergence can be interpreted in much the same way as the conditions which guarantee the stability of a discrete-time system. In other words, if the z-plane diagram of a sequence contains poles, then *all* of the poles, real and complex, must be contained by the unit circle. There is, of course no such restriction on the zeros of a signal. If, as is often the case, zeros occur on the unit circle, these will give rise to spectral nulls or – in the case of frequency response functions – to transmission zeros at certain frequencies.

Worked Example 7.2 Find an expression for the DTFT of the infinite sequence defined by

$$x[n] = \alpha^n;\ n \geq 0$$
$$= 0;\quad n < 0.$$

Solution: It was shown in Chapter 5 that a sequence of the type $x[n] = 1, \alpha, \alpha^2, \alpha^3, \ldots$ is characterized by the z-transform

$$X(z) = \frac{z}{z - \alpha}. \quad (7.16)$$

$X(z)$ has a pole at the point $z = \alpha$ in the z-plane. The pole will lie within the unit circle, provided that $|\alpha| < 1$. Subject to this condition being satisfied, $x[n]$ will be absolutely summable and its DTFT will be given by the convergent series

$$X(e^{j\omega T}) = 1 + \alpha e^{-j\omega T} + \alpha^2 e^{-j2\omega T} + \alpha^3 e^{-j3\omega T} + \ldots$$

The discussion linking Equations 5.18 and 5.21 in Chapter 5 deals with the relationship between a rational function and its series expansion.

We can interpret this series as the series expansion of the following closed-form expression which is obtained by setting $z = e^{j\omega T}$ in Equation 7.16:

$$X(e^{j\omega T}) = \frac{e^{j\omega T}}{e^{j\omega T} - \alpha}.$$

To calculate the magnitude and phase of the spectrum, we work with the closed form of $X(e^{j\omega T})$ and follow the development in Worked Example 5.12 in Chapter 5, giving:

$$|X(e^{j\omega T})| = \frac{1}{\sqrt{(1 - 2\alpha\cos\omega T + \alpha^2)}}$$

$$\Phi = -\tan^{-1}\left(\frac{\alpha\sin\omega T}{1 - \alpha\cos\omega T}\right).$$

The spectral magnitude and phase in the previous example are plotted in Figure 7.4 for the case $\alpha = 0.8$. The behaviour of the spectral magnitude for other values of α can be inferred by inspecting the pole-zero diagram of the z-transform $X(z)$ which is identical to that of the transfer function $H(z)$ describ-

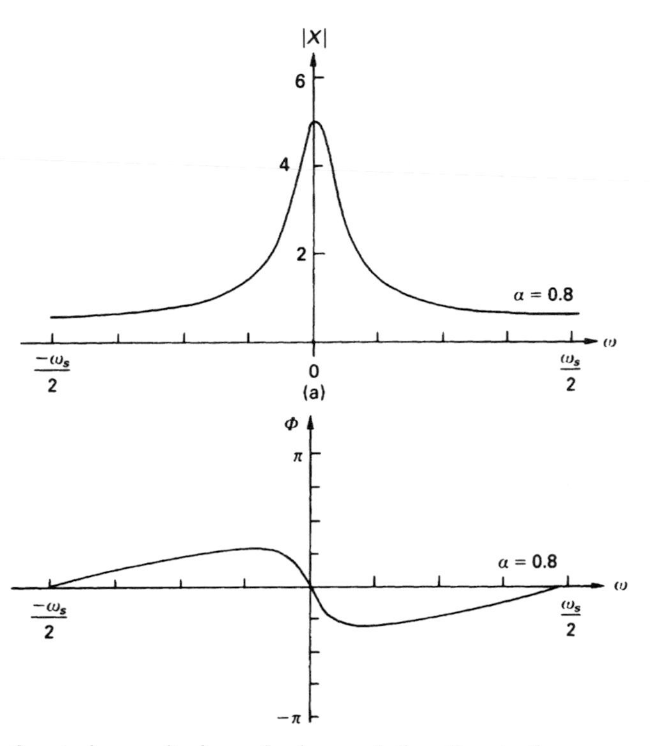

Fig. 7.4 Spectral magnitude and phase of the discrete-time sequence of Worked Example 7.2, with $\alpha = 0.8$.

ing a simple recursive discrete-time system. Reference to Figure 5.15 and the related discussion in Chapter 5 shows that as the value of α is reduced, the spectral content of $x[n]$ becomes more uniformly distributed in the frequency range $|\omega| < \omega_s/2$. Conversely, as α is increased, the pole at $z = \alpha$ moves closer to the point $z = 1$ so that the spectral content of $x[n]$ becomes more concentrated at low frequencies. It should be remembered, however, that if the pole is placed on the unit circle or outside the unit circle, the DTFT is no longer defined and the sequence $x[n]$ cannot be described in terms of its spectrum.

If $\alpha = 1$, the sequence $x[n] = 1$ for all n. If $|\alpha| > 1$ the values of $x[n]$ will increase indefinitely with n.

Exercise 7.2

Which of the following sequences possesses a convergent DTFT?

a) $x[n] = \delta[n]$; b) $x[n] = u[n]$; c) $x[n] = 2^n u[n]$;
d) $x[n] = u[n](0.5)^n \cos(n/2)$; e) $x[n] = (0.2)^{|n|}$, $-\infty < n < \infty$.

Multiplication by the unit-step sequence $u[n]$ defines a sequence to be zero for all $n < 0$.

Symmetry properties of the DTFT

The DTFT has symmetry properties analogous to those of the continuous-time Fourier transform described in Chapter 3. For instance, in each of the examples considered so far we have seen that the spectral magnitude $|X(e^{j\omega T})|$ is an even function of ω:

$$|X(e^{j\omega T})| = |X(e^{-j\omega T})| \tag{7.17}$$

and that the phase is an odd function of ω:

$$\Phi(\omega) = -\Phi(-\omega). \tag{7.18}$$

It can be shown that these are general properties of the DTFT applied to any real-valued sequence $x[n]$. Using a similar approach to that given in connection with Equations 3.43 and 3.44 in Chapter 3, we can further establish that the DTFT of a real-valued sequence has the property of conjugate symmetry, that is

$$X(e^{-j\omega T}) = X^*(e^{j\omega T}) \tag{7.19}$$

where * denotes the complex conjugate.

We can sometimes use these properties to advantage when calculating spectra and plotting graphs since if we know $|X(e^{j\omega T})|$ and $\Phi(\omega)$ for $0 \leqslant \omega \leqslant \omega_s/2$, we also know these functions for $-\omega_s/2 \leqslant \omega \leqslant 0$. Further simplification can also follow by using the properties reviewed in the following exercise.

Exercise 7.3 The first part of this exercise illustrates properties of the DTFT which depend on the symmetry of the original sequence.

a) A real even sequence $x_E[n]$ and a real odd sequence $x_O[n]$ are defined by

$$x_E[n] = x_E[-n]; \qquad x_O[n] = -x_O[-n]$$

Establish the following properties of their spectral models:

i) $X_E(e^{j\omega T}) = x[0] + 2\sum_{n=1}^{\infty} x[n]\cos n\omega T$

ii) $X_O(e^{j\omega T}) = -2j\sum_{n=1}^{\infty} x[n]\sin n\omega T.$

b) Show that if $x[n]$ has a spectrum $X(e^{j\omega T})$ the delayed sequence $x[n-k]$ will have the spectrum $X(e^{j\omega T})e^{-jk\omega T}$.

The inverse DTFT

We have calculated forward transforms of sequences by evaluating a complex exponential series:

$$X(e^{j\omega T}) = x[0] + x[1]e^{-j\omega T} + x[2]e^{-2j\omega T} + \ldots$$

We assume that the sampling time T is constant so that, for a given value of n, the complex exponential component $e^{-jn\omega T}$ is a continuous function of frequency.

Each term in the series comprises the product of a coefficient, given by a sequence value $x[n]$, and a complex exponential component of the form $e^{-jn\omega T}$. What we have, in fact, is a Fourier series description of the periodic frequency function $X(e^{j\omega T})$. The periodicity is inherent in the frequency-dependent harmonic components $e^{-jn\omega T}$, which share a common period of length $\omega_s = 2\pi/T$, measured along the frequency axis.

If signal spectra always appeared in the series form, then it would be no problem at all to *invert* a spectrum to find the values of the corresponding discrete-time sequence. We would simply read off the coefficients of the complex components $e^{-jn\omega T}$ for $n = 0, 1, 2, \ldots$ and set these equal to $x[0]$, $x[1]$, $x[2]$, ... But, as we have seen, spectral models and frequency response functions are often expressed in closed form, involving a mixture of exponential and trigonometrical functions. Furthermore, these functions are often multiplied together, for example when finding the output spectrum in a linear discrete-time system:

$$Y(e^{j\omega T}) = H(e^{j\omega T}) \times X(e^{j\omega T}).$$

We cannot always depend on the series form of $Y(e^{j\omega T})$ being readily available, but we can be sure that $Y(e^{j\omega T})$ has the same period, ω_s, as $H(e^{j\omega T})$ and $X(e^{j\omega T})$. In principle, then, we can analyse the periodic spectrum using Fourier techniques, with the aim of finding the values of the steady-state output sequence $y[n]$ represented by the coefficients $y[n]$ of the Fourier series

$$Y(e^{j\omega T}) = y[0] + y[1]e^{-j\omega T} + y[2]e^{-2j\omega T} + \ldots$$

The decomposition of $Y(e^{j\omega T})$ can be achieved by an appropriate modification to the Fourier analysis techniques described in Chapter 6 in relation to periodic continuous-time signals. We set up a 'sifting' integral by first multiplying the periodic function $Y(e^{j\omega T})$ by $e^{jn\omega T}$. The product is then averaged over a single period of length ω_s. In this way, we arrive at a formal definition of the inverse DTFT:

$$y[n] = \frac{1}{\omega_s}\int_0^{\omega_s} Y(e^{j\omega T})e^{jn\omega T}\,d\omega. \qquad (7.20)$$

We shall give an example of the application of Equation 7.20 later, in Worked Example 7.5.

Frequency scaling and normalization

An alternative form of the inverse transform may be obtained by substituting $\theta = \omega T$ in Equation 7.20. We set $d\omega = d\theta/T$ and evaluate the integral over the range $\theta = 0$ to $\theta = \omega_s T = 2\pi$, giving

$$y[n] = \frac{1}{2\pi}\int_0^{2\pi} Y(e^{j\theta})e^{jn\theta}\,d\theta. \qquad (7.21)$$

The spectral function $Y(e^{j\theta})$ represents the value of the signal z-transform $Y(z)$ at the point $e^{j\theta}$ on the unit circle. As θ varies between $\theta = 0$ and $\theta = 2\pi$ the values of $Y(e^{j\theta})$ correspond to those of $Y(e^{j\omega T})$ taken over $\omega = 0$ to $\omega = 2\pi/T$, as illustrated in Figure 7.5. Since $Y(e^{j\omega T})$ is periodic in frequency, with period ω_s, $Y(e^{j\theta})$ is periodic in θ, with a period of 2π.

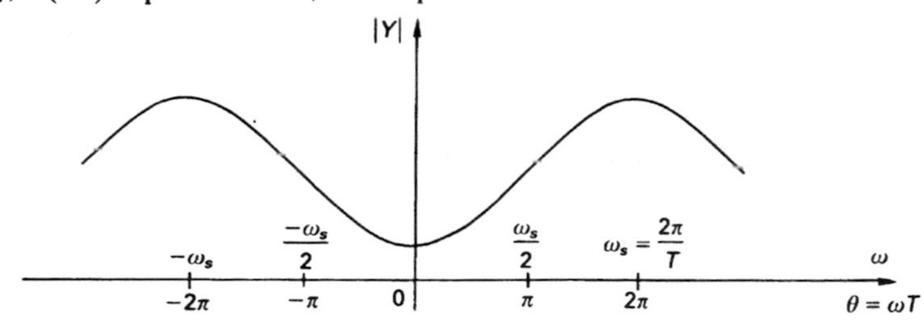

Fig. 7.5 Rescaling the frequency axis when $\theta = \omega T$.

Worked Example 7.3

Show that $Y(e^{j\theta})$ is 2π-periodic.

Solution: If we replace θ with $\theta + 2\pi$, we obtain

$$Y(e^{j(\theta+2\pi)}) = \sum_{n=-\infty}^{\infty} x[n]e^{-jn\theta}e^{-j2\pi n}$$

which is equal to $Y(e^{j\theta})$, since $e^{-j2\pi n} = 1$. Repeating the calculation for $\theta + M\pi$, where M is any integer we can show that $Y(e^{j(\theta+2M\pi)}) = Y(e^{j\theta})$, confirming that $Y(e^{j\theta})$ is 2π-periodic.

We shall see later that the '$e^{j\theta}$' form of the spectrum is the preferred form when we carry out numerical computations of spectra. Now, as the defining Equations 7.9 and 7.11 make clear, the derivation of the DTFT depends only on the *product* of the frequency ω and the sampling interval T. By making the substitution $\theta = \omega T$, we can calculate a spectrum directly in terms of $e^{j\theta}$;

$$X(e^{j\theta}) = \sum_{n=-\infty}^{\infty} x[n]e^{-jn\theta}. \tag{7.22}$$

Worked Example 7.4

Find the spectrum $X(e^{j\theta})$ of the finite sequence shown opposite and sketch its spectral magnitude over the range $-\pi \leqslant \theta \leqslant \pi$.

What will be the frequency of the spectral null in the range $0 < \omega < \omega_s/2$ if the sampling interval is 125 μs?

Solution: $x[n] = 1$ for $n = 0, 1, 2$ and is zero for all other n. Its spectrum is given by the finite summation

$$X(e^{j\theta}) = \sum_{n=0}^{2} e^{-jn\theta} = 1 + e^{-j\theta} + e^{-2j\theta}$$

$$= e^{-j\theta}(e^{j\theta} + 1 + e^{-j\theta})$$

$$= e^{-j\theta}(1 + 2\cos\theta).$$

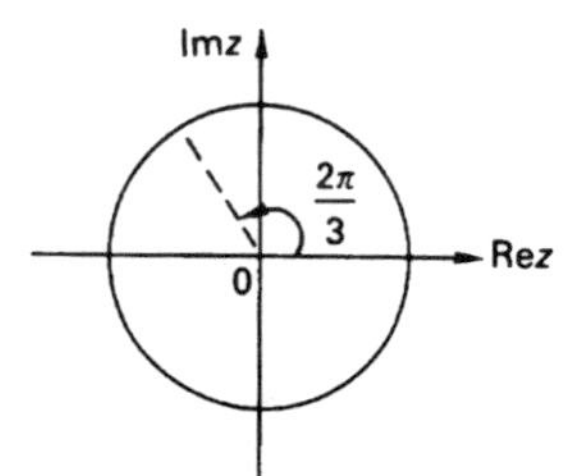

The zeros of $X(e^{j\theta})$ correspond to the zeros of the z-transform $X(z)$ on the unit circle, as shown here.

The graph of the spectral magnitude $|X(e^{j\theta})| = |1 + 2\cos\theta|$ is shown in Figure 7.6. The spectral nulls in the range $-\pi < \theta < \pi$ correspond to the zero values of $(1 + 2\cos\theta)$ which occur at $\theta = \pm 2\pi/3$ radians.

To obtain the frequency of the spectral null for a given sampling interval T, we rescale the horizontal axis by setting $\theta = \omega T$. The point $\theta = \pi$ then corresponds to the frequency $\omega = \pi/T = \omega_s/2$.

If $T = 125\,\mu s$ then $f_s = 1/T = 8\,kHz$. The frequency of the spectral null at $\theta = 2\pi/3$ is therefore $4\,kHz \times 2/3 = 2.66\,kHz$.

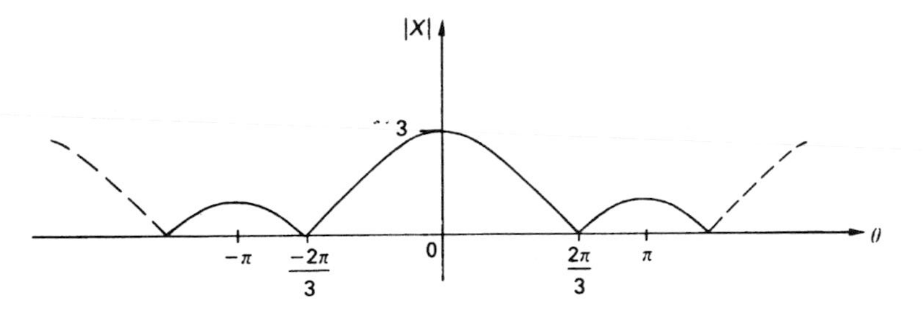

Fig. 7.6 Spectral magnitude $|X(e^{j\theta})|$ for the discrete-time sequence of Worked Example 7.4.

If we use Equation 7.22 to transform a sequence $x[n]$ to the frequency domain we can modify the frequency content of $x[n]$ by operating directly on the spectrum $X(e^{j\theta})$. The modified spectrum $Y(e^{j\theta})$ can then be inverted using the inverse transform, Equation 7.21, all without making reference to the sampling frequency ω_s or the sampling interval T. This turns out to be important when we carry out spectral analysis and filtering of data sequences using a digital computer since the only input required is the set of ordered numerical values provided by the sequence $x[n]$. Similarly, the result of the calculation is a string of numbers representing the values of the output sequence $y[n]$. Any information we have regarding the sampling frequency or notional sample spacing may be used to establish time and frequency scales and assist in interpreting and comparing results. But, as a rule, such information is usually brought into play after the main calculation has been carried out.

When we use the $e^{j\theta}$ form of the spectrum we are expressing a sequence in terms of components of the form $e^{jn\theta}$. The appropriate form of the frequency-response function is then $H(e^{j\theta})$ which can be compared directly with a spectrum $X(e^{j\theta})$.

We can program a computer to evaluate the convolution sum $y[n] = h[n] * x[n]$ without having any knowledge of the sampling time T. Similarly, if we use the $e^{j\theta}$ form of the spectrum, the calculation of the steady-state output sequence $y[n]$ can be carried through without prior knowledge of the sample spacing.

Signal sampling and aliasing

We first saw in Chapter 1 that a continuous-time signal $x(t)$ can be completely represented by a sequence of sample values $x[n]$, taken at regular intervals, T, along the time axis. In practice, the waveform and spectrum of an information-bearing signal, such as a speech or a music signal, may be highly complicated. Nevertheless, the transfer of signal information from the continuous-time domain to a set of samples can be accomplished with relative ease because effective sampling does not depend on detailed prior knowledge of a signal. As explained in Chapter 1, it is sufficient to know that the signal contains frequency components up to some maximum bandwidth frequency, f_B. Provided that the sampling frequency f_s is chosen so that $f_s > 2f_B$, in accordance with the sampling rule, the waveform of the original signal can be recovered without distortion by interpolating its sample values $x[n]$.

It is usual practice to pass a signal through a low-pass filter prior to sampling. The filter has a well-defined cut-off frequency and is usually known as an anti-aliasing filter.

The immediate implication here is that we can store samples of measurement data on computer tape or disk and retrieve them at any later date for the purposes of displaying and comparing waveforms. But why stop here? All of the operations that we require of a discrete-time system – delay, addition and scaling by a constant and summation of sequences – can be programmed into a digital computer with the aim of processing the data sequence in some desired fashion. We might think initially in terms of time-domain filtering operations.

But as we shall see, the same computer could also be used to calculate forward and inverse Fourier transforms, making it possible to view all operations on a signal from a frequency-domain as well as a time-domain perspective. Once in the frequency domain we know that the operation of any linear processing system can be described in terms of its filtering effect on the frequency components of an input signal. It is then a fairly routine task to adjust the magnitude and phase of a signal spectrum by multiplication with a frequency-response function, before transforming back to the time domain to compare the processed signal with the original input.

The fact that all of the above can be accomplished using a personal computer loaded with proprietary software makes it all the more essential that we appreciate the limitations as well as the advantages of working with signal samples. Since our main concern here is with frequency-domain models we shall consider first of all the relationship between the spectrum of a continuous-time signal $x(t)$:

$$X(j\omega) = \int_{-\infty}^{\infty} x(t)\,\mathrm{e}^{-j\omega t}\,\mathrm{d}t \tag{7.23}$$

and the spectrum of its sample sequence $x[n]$, given by the DTFT:

We are assuming here that $x(t)$ and, hence, its samples $x[n]$ are finite-energy signals.

$$X(\mathrm{e}^{j\omega T}) = \sum_{n=-\infty}^{\infty} x[n]\,\mathrm{e}^{-jn\omega T}. \tag{7.24}$$

We shall assume for practical purposes that $x(t)$ is a bandlimited signal with bandwidth f_B. In this case $X(j\omega)$ will be a bandlimited function, taking negligible values for $|\omega| > \omega_B$, where $\omega_B = 2\pi f_B$. If the sampling rule is obeyed, then $X(\mathrm{e}^{j\omega T})$ will also be bandlimited, as in Figure 7.7. Notice that within the fundamental interval, $0 \leqslant \omega \leqslant \omega_s$, all of the significant values of $X(\mathrm{e}^{j\omega T})$ are contained within the bandwidth ω_B. We also have $\omega_B < \omega_s/2$, in accordance with the sampling rule.

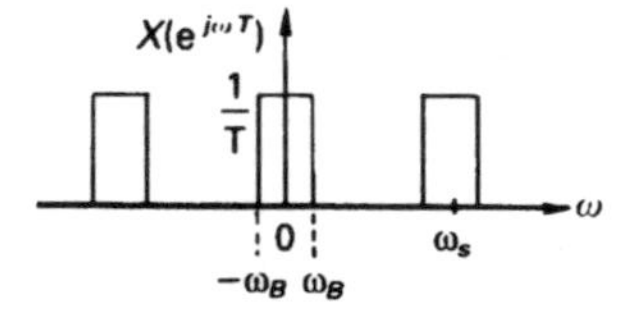

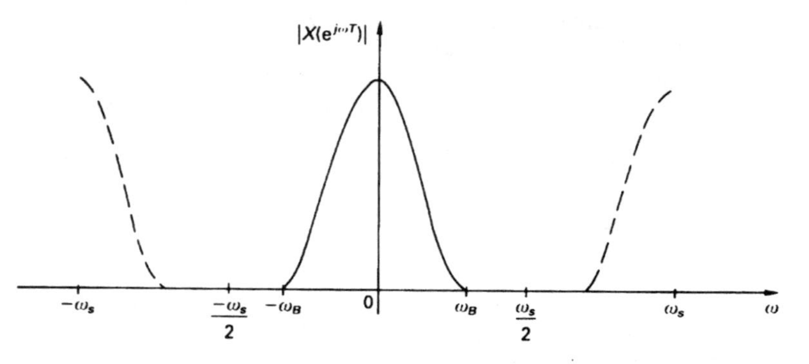

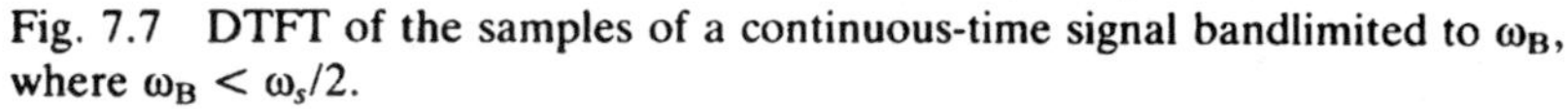

Fig. 7.7 DTFT of the samples of a continuous-time signal bandlimited to ω_B, where $\omega_B < \omega_s/2$.

Worked Example 7.5 Find the sample sequence $x[n]$ associated with the spectrum $X(\mathrm{e}^{j\omega T})$ shown opposite.

Solution: We interpret $X(e^{j\omega T})$ as the spectrum of a discrete-time signal strictly bandlimited to the frequency range $|\omega| < \omega_B$. $X(e^{j\omega T})$ is a real function in this example, completely specified by its behaviour for $|\omega| < \omega_s/2$. We have:

$$X(e^{j\omega T}) = 1/T, \quad \text{for } |\omega| < \omega_B$$
$$= 0, \quad \text{for } \omega_B < |\omega| < \omega_s/2.$$

If we interpret $X(e^{j\omega T})$ as a frequency-response function, then the related discrete-time system will have the characteristics of a highly-idealized low-pass filter, cutting off abruptly at the frequency ω_B.

$x[n]$ is given by taking the inverse DTFT of $X(e^{j\omega T})$. We therefore multiply $X(e^{j\omega T})$ by $e^{jn\omega T}$ and average the product over one period of length ω_s, giving:

$$x[n] = \frac{1}{\omega_s T}\int_{-\omega_s/2}^{\omega_s/2} e^{jn\omega T}\,d\omega = \frac{1}{2\pi}\int_{-\omega_B}^{\omega_B} e^{jn\omega T}\,d\omega.$$

Remember that we have dealt with this integral before, when finding the spectrum of a rectangular pulse signal.

After taking the integral and rearranging, we obtain:

$$x[n] = \frac{\omega_B}{\pi}\frac{\sin n\omega_B T}{n\omega_B T}.$$

which we can interpret as the samples of a continuous-time '$\sin x/x$' function of the type that we originally met in Chapter 3.

The continuous-time '$\sin x/x$' function from the previous example is shown in Figure 7.8 together with its spectrum $X(j\omega)$. If we now compare $X(j\omega)$ with $X(e^{j\omega T})$ illustrated in the lower half of the figure, we see that $X(e^{j\omega T})$ can be regarded as the *periodic extension* of the scaled spectrum $X(j\omega)/T$. This is in line with the approach developed in Chapter 6, where we modelled a periodic time variation, and means that $X(e^{j\omega T})$ is given by adding together shifted versions of $X(j\omega)/T$ centred on the frequencies $\pm\omega_s$, $\pm 2\omega_s$, $\pm 3\omega_s$ and so on.

To relate $x(t)$ and $X(j\omega)$, all we have to do is to take the inverse Fourier transform of $X(j\omega)$. We thus obtain

$$x(t) = \frac{1}{2\pi}\int_{-\omega_B}^{\omega_B} e^{j\omega t}\,d\omega = \frac{\omega_B}{\pi}\frac{\sin \omega_B t}{\omega_B t}$$

Now, in this example we have $\omega_B < \omega_s/2$, which means that the sampling frequency is greater than twice the signal bandwidth. As a result, the shifted versions of $X(j\omega)$ are well-separated along the frequency axis and we can state:

$$X(e^{j\omega T}) = X(j\omega)/T, \text{ for } |\omega| < \omega_s/2 \tag{7.25}$$

In other words, apart from a scaling factor, the spectrum of the signal $x(t)$ and the spectrum of its samples $x[n]$ are identical within the low-frequency range $|\omega| < \omega_s/2$. However, this is only the case because $x(t)$ has been sampled in

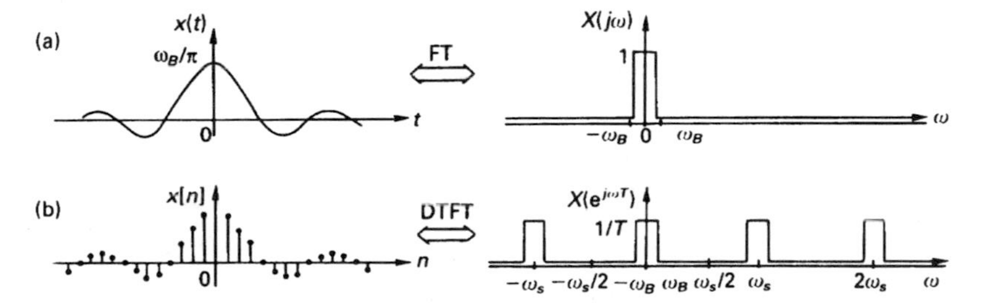

Fig. 7.8 Continuous spectra of (a) a continuous-time '$\sin x/x$' signal and (b) a sampled '$\sin x/x$' signal.

accordance with the sampling theorem. To see what happens otherwise we shall now consider the continuous signal $x(t)$ shown in Figure 7.9 with its spectrum $X(j\omega)$. To find the spectrum of its samples $x[n]$, we proceed as before and plot shifted versions of $X(j\omega)$ along the frequency axis, spaced at intervals ω_s and scaled by $1/T$. The superposition of the shifted spectra then gives us the periodic spectrum $X(e^{j\omega T})$. If the sampling theorem has been obeyed as in Figure 7.9b, then Equation 7.25 will apply and the two spectra will have similar form for $|\omega| < \omega_s/2$. If, on the other hand, the sampling frequency is too low, as in Figure 7.9c, the shifted versions of $X(j\omega)/T$ will overlap and $X(e^{j\omega T})$ takes a quite different form. What we have in fact is a graphical illustration of the phenomenon of aliasing whereby high-frequency components of $x(t)$ take on the identity of low-frequency components and appear in $X(e^{j\omega T})$ for $|\omega| < \omega_s/2$.

In this context, 'high frequency' means the components in the 'tails' of $X(j\omega)$ for frequencies $|\omega| > \omega_s/2$.

We know that if a signal has not been sampled properly, following the sampling theorem, then it will be impossible to recover its waveform by interpolating its samples. Now, in view of Figure 7.9, we see that it will also be impossible to deduce the form of its spectrum, owing to the distortion introduced by the aliased components. When the sampling theorem is obeyed on the other hand, complete information about the spectrum of $x(t)$ is retained in its samples and we can calculate values of $X(j\omega)$ by evaluating the Fourier sum

This expression tells how to find $X(j\omega)$ in terms of $X(e^{j\omega T})$ and involves the factor T suggested by Equation 7.25.

$$X(j\omega) = TX(e^{j\omega T}) = T \sum_{n=-\infty}^{\infty} x[n]\,e^{-jn\omega T} \tag{7.26}$$

for frequencies in the range $|\omega| < \omega_s/2$.

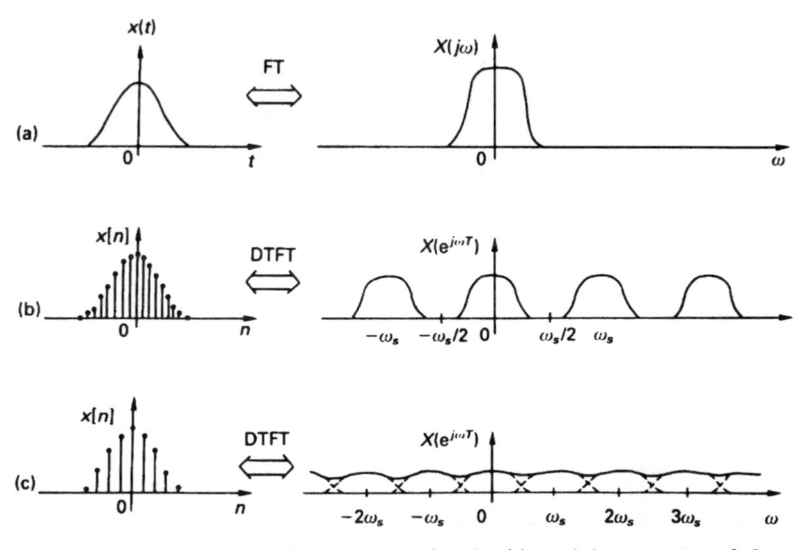

Fig. 7.9 Spectra of a continuous-time signal $x(t)$ and its samples $x[n]$ (a) original signal; (b) sampling theorem obeyed; (c) effect of aliasing when the sampling rate is too low.

Frequency resolution

We have seen that the spectrum of a finite-energy signal can be related directly to the spectrum of its samples, provided that the signal $x(t)$ is bandlimited and sampled in accordance with the sampling rule. Suppose now that $x(t)$ is an experimental signal. There is no difficulty in ensuring that $x(t)$ is bandlimited; all we need is an anti-aliasing filter with sufficiently sharp cut-off. But if $x(t)$ were truly aperiodic and, thus, defined over the entire time axis it would be quite impracticable to take an infinite set of samples. The only feasible approach is to *truncate* the signal to an interval of finite length T_o as shown in Figure 7.10 and derive from it a finite-length sequence $x[n]$.

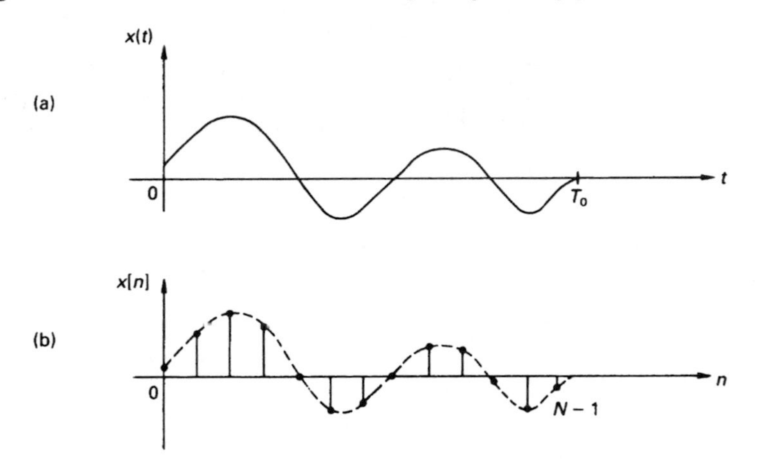

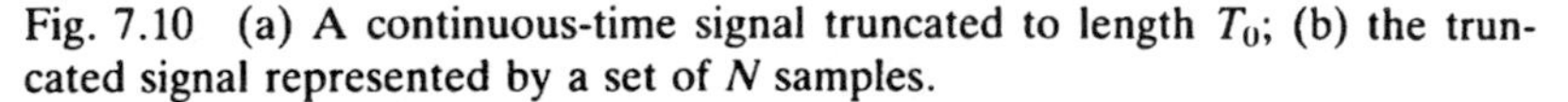
Fig. 7.10 (a) A continuous-time signal truncated to length T_0; (b) the truncated signal represented by a set of N samples.

The spectrum of $x[n]$ is now given by the finite summation

$$X(e^{j\omega T}) = \sum_{n=0}^{N-1} x[n]\, e^{-jn\omega T} \tag{7.27}$$

which can be inserted in Equation 7.26 to give an approximation to the spectrum $X(j\omega)$ of $x(t)$:

$$X(j\omega) \approx X(e^{j\omega T})/T. \tag{7.28}$$

We indicate an approximation here rather than an equality because the finite-length sequence $x[n]$ and, hence, its spectrum $X(e^{j\omega T})$ contain no information about $x(t)$ outside of the truncation interval. The missing information is, in fact, provided by all the frequency components of $x(t)$ with periods greater than T_0. As soon as the truncation 'window' was imposed we effectively excluded all these components with frequencies less than $1/T_0$ from subsequent analysis.

In practice, this loss of low-frequency information makes it impossible to distinguish between spectral components of $x(t)$ spaced more closely than $\delta f = 1/T_0$. We say that the *frequency resolution* is finite and equal to the reciprocal of the record length:

frequency resolution, $\delta f = 1/T_0$.

The importance of working with an appropriate frequency resolution is illustrated by the following example. First, in Figure 7.11a, we show a portion of the spectrum $X(j\omega)$ of $x(t)$ before truncation. Notice that $X(j\omega)$ contains a spectral feature consisting of two peaks separated by an amount $\Delta\omega = 2\pi\,\Delta f$. Now, it is precisely this sort of spectral detail which might be obscured when $x(t)$ is truncated to a finite length T_0. If we consider such detail to be important then we must arrange for the overall spectral resolution $\delta f = 1/T_0$ to be less than the critical spacing Δf. In other words, we must choose T_0 so that $T_0 > 1/\Delta f$. When this criterion is met, as in Figure 7.11b, we have sufficient resolution to ensure that the calculated spectrum $X(e^{j\omega T})/T$ reproduces the fine detail of $X(j\omega)$. When the truncation length is made too short, as in Figure 7.11c the calculated results provide a much smoother version of $X(j\omega)$. Besides blurring the spectral detail, we see that the use of a short truncation interval has also resulted in a significant broadening of the calculated spectrum relative to $X(j\omega)$.

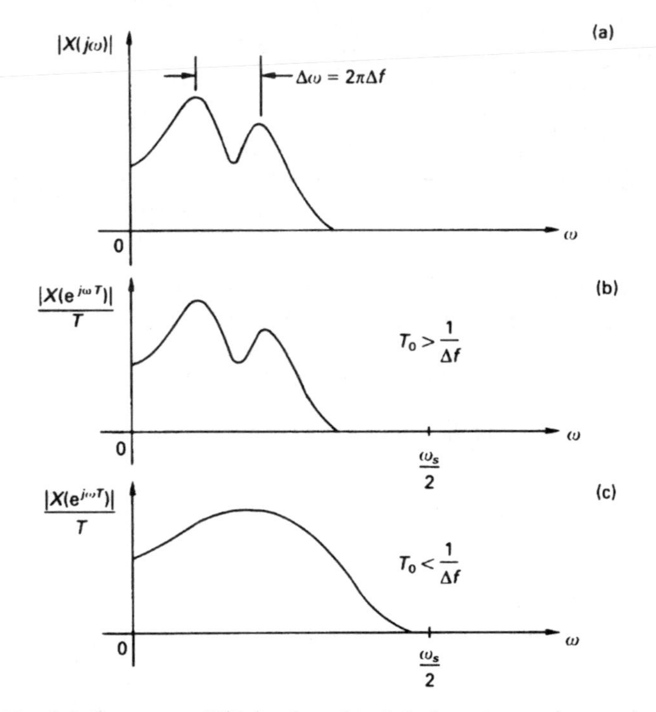

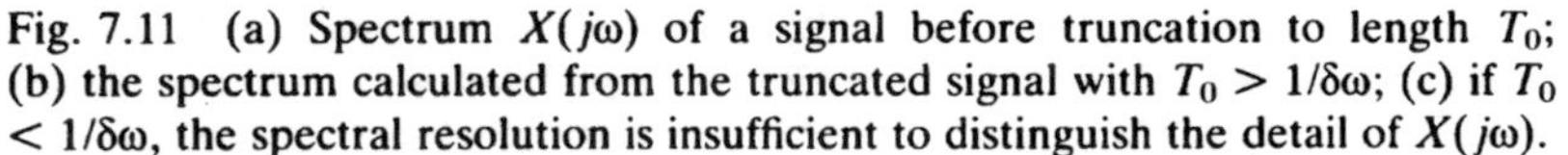

Fig. 7.11 (a) Spectrum $X(j\omega)$ of a signal before truncation to length T_0; (b) the spectrum calculated from the truncated signal with $T_0 > 1/\delta\omega$; (c) if $T_0 < 1/\delta\omega$, the spectral resolution is insufficient to distinguish the detail of $X(j\omega)$.

Worked Example 7.6 A set of samples are taken from a transducer output signal $x(t)$ with a view to finding the spectrum of the signal. The signal bandwidth is limited by the response time of the transducer and estimated to be 2 kHz.

The spectrum is required with a resolution of 10 Hz or less. Calculate a) the minimum record length; b) the maximum allowable sampling time; c) the minimum number of signal samples required.

The use of transducers is dealt with in *Transducers and Interfacing*: B.R. Bannister and D.G. Whitehead, Tutorial Guides in Electronic Engineering No. 7, Chapman and Hall.

Solution:

a) To achieve a frequency resolution of 10 Hz or less, we require $T_0 \geq 1/10$ s. The minimum record length is therefore 100 ms.
b) The maximum possible sampling spacing is given by sampling at the lowest allowed frequency, $f_s = 2f_B$. Since $f_B = 2$ kHz, the maximum sampling interval is $T = 0.25$ ms.
c) The number of samples required is T_0/T or f_sT_0. The minimum number is found using $f_s = 4$ kHz or $T = 0.25$ ms, giving $N = 400$.

It is not considered good practice to sample a signal at the frequency $f_s = 2B$. Strictly speaking, the sampling frequency should be chosen so that $f_s > 2B$ to avoid aliasing errors.

Since we cannot avoid truncating a signal we must always accept some loss of resolution accompanied by broadening and additional distortion of the spectrum not shown in Figure 7.11. In many cases, the distortion can be kept to some minimum acceptable level by the use of special *windowing* techniques. It should be emphasised, however, that such techniques cannot improve the spectral resolution beyond the limit set by the length of the truncation interval.

We remarked on the need for signal truncation in Chapter 2. A marginal comment on page 31 cites two references which deal with the topic of windowing.

Spectral sampling

Truncating a signal can have undesirable effects, but it makes for a relatively easy task when we evaluate the spectrum of the signal samples, $X(e^{j\omega T})$. The most obvious benefit is that we always work with a finite summation as indicated by Equation 7.27.

Further to this, the truncated signal provides a spectrum $X(e^{j\omega T})$ that varies smoothly over any frequency interval comparable with the frequency resolution $\delta\omega = 2\pi\delta f = 2\pi/T_0$. In this case, there is little to be gained by computing values of $X(e^{j\omega T})$ for frequencies spaced more closely than $\delta\omega$. Provided that we choose frequencies spaced by no more than $\delta\omega$, we should have sufficient information to reconstruct the graph of $X(e^{j\omega T})$ by interpolating the spectral samples.

Since $X(e^{j\omega T})$ is a periodic function of ω, we can define its behaviour by taking a set of samples spaced at intervals $\delta\omega = 2\pi/T_0$ over the range $0 \leq \omega < \omega_s$ as indicated in Figure 7.12. The number of spectral values required is

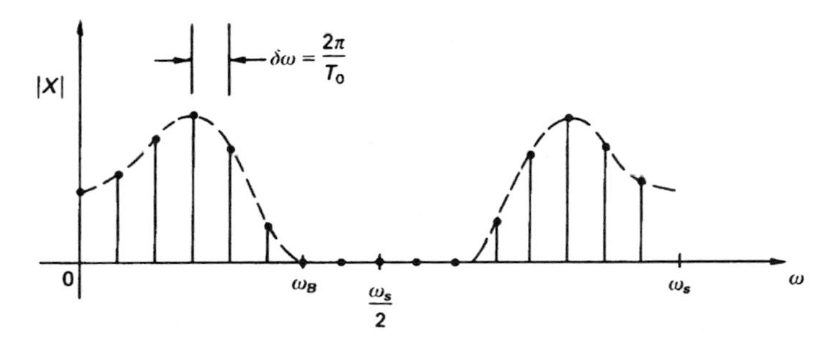

Fig. 7.12 Taking spectral samples at a spacing $\delta\omega = 2\pi/T_0$.

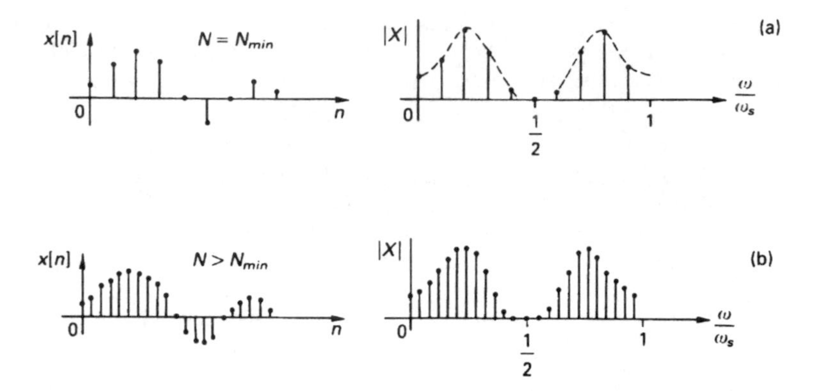

Fig. 7.13 (a) A finite-duration signal sampled at the minimum rate; (b) taking more samples over the same interval gives better interpolation but does not reveal additional spectral detail.

We have concluded that a properly sampled signal can be represented by N time domain values and by the same number, N, spectral values.

$$\omega_s/\delta\omega = f_s T_0 = T_0/T \tag{7.28}$$

which is equal to the number N of time-domain samples representing the truncated signal.

If we now refer to Worked Example 7.6, we see that the minimum number of samples required is $N_{\text{min}} = 2f_B T_0$, obtained by sampling a bandlimited signal at the lowest frequency permitted by the sampling rule. We may, of course choose $N > N_{\text{min}}$. But if the record length remains fixed, such *oversampling* does not reveal additional spectral detail. As Figure 7.13 shows, the only effect is to fill in the gaps along the frequency axis. This provides better interpolation of $X(e^{j\omega T})$ but it cannot recreate the spectral information that was lost when the truncation 'window' was applied to $x(t)$ at the start of the analysis. If better spectral resolution is required, then the only solution is to use a longer record length.

Since the sampling frequencies are different in Figures 7.13a and 7.13b, the frequency axes have been scaled with respect to the sampling frequency, for ease of comparison.

With these considerations in mind we shall now look in more detail at the calculation of spectral samples and aim to establish the relationship between the values of a finite-length sequence and the sample values of its spectrum.

The discrete Fourier transform

We begin by defining a finite-length sequence $x[n]$ with N distinct values:

$$x[n] = x_n,\ n = 0, 1, 2, \ldots, N-1. \tag{7.29}$$

If we wish to calculate the spectrum of $x[n]$ it is usually easiest to work with the '$e^{j\theta}$' form, given by the finite sum:

$$X(e^{j\theta}) = \sum_{n=0}^{N-1} x[n]\,e^{-jn\theta}. \tag{7.30}$$

It will be recalled that $X(e^{j\theta})$ is obtained by setting $\theta = \omega T$ in Equation 7.11, defining the forward DTFT.

Since $x[n]$ is a finite-length sequence, we can represent $X(e^{j\theta})$ by a set of N spectral values equally spaced over the range $-\pi \leqslant \theta \leqslant \pi$ or $0 \leqslant \theta \leqslant 2\pi$.

In practical applications, a digital computer is almost always used for this purpose. We thus envisage a computer program that transforms one finite set of data, $x[n]$, into a second finite set $X[k]$, denoting sample values of $X(e^{j\theta})$ taken over some interval of length 2π.

Since $X(e^{j\theta})$ is periodic it is sufficient to define its behaviour over a single period. However, the computer cannot work with a continuous variable θ, so we must be content with representing $X(e^{j\theta})$ by a set of sample values.

When this procedure is defined more rigorously, we find that the relationship between the values of $x[n]$ and $X[k]$ can be described formally by means of the *discrete Fourier transform* or DFT. The DFT is by far the most widely used of all the transformations we have considered. This is due to the fact that it can be executed rapidly and with the minimum of computational effort using a special class of algorithm known as the Fast Fourier Transform (FFT). Since the FFT was first described in 1965 it has played an increasingly important role in almost all aspects of digital signal processing, including spectrum analysis, digital filter design and fast convolution for FIR systems, besides a host of other applications beyond the scope of our present treatment.

J.W. Cooley and J.W. Tukey were the first to publish details of a fast algorithm for the machine computation of the DFT.

Calculating spectral samples

We shall follow the sampling scheme illustrated in Figure 7.14 and calculate N values of the spectrum $X(e^{j\theta})$ at equidistant points on the interval $\theta = 0$ to $\theta = 2\pi$. The spectral spacing is thus $2\pi/N$.

Remember that the number of spectral samples taken is equal to the length, N, of the sequence $x[n]$.

Starting with $\theta = 0$, we generate the sequence:

$$X[0] = X(e^{j0}),\ X[1] = X(e^{j2\pi/N}), \ldots X[k] = X(e^{j2\pi k/N})$$

and so on, up to the final value:

$$X[N-1] = X(e^{j2\pi(N-1)/N}).$$

The N values of $e^{j\theta}$ used in the computation define a set of points equally spaced around the unit circle as shown in Figure 7.15 overleaf. This diagram reminds us that we can regard the $X[k]$ either as samples of the DTFT $X(e^{j\theta})$ or as samples of the z-transform $X(z)$, taken at the points $z = e^{j2\pi k/N}$, $k = 0, 1, \ldots, N-1$:

$$\begin{aligned} X[k] &= X(e^{j\theta})|_{\theta=2\pi k/N} \\ &= X(z)|_{z=e^{j2\pi k/N}};\ k = 0, 1, 2, \ldots, N-1. \end{aligned} \qquad (7.31)$$

For a given value of N, the complex exponential $e^{j2\pi k/N}$ is a function of the integer variable k. Figure 7.15 shows that the sequence $w[k] = e^{j2\pi k/N}$ is a periodic sequence which repeats its values whenever k is increased or decreased by N, i.e. $w[k] = w[k \pm mN]$, where m is any integer.

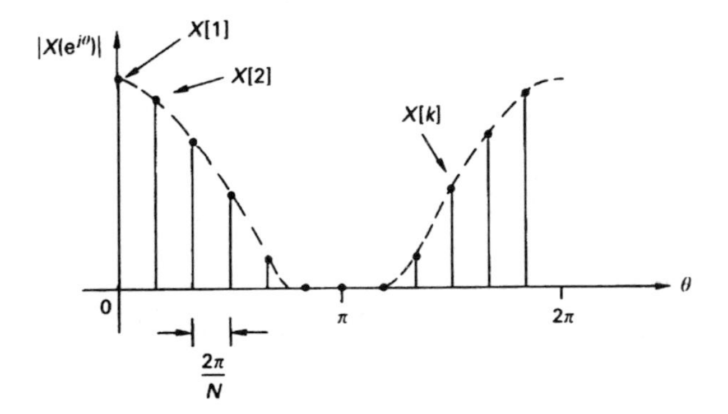

Fig. 7.14 Calculating N spectral values $X[k]$ spaced at intervals $2\pi/N$.

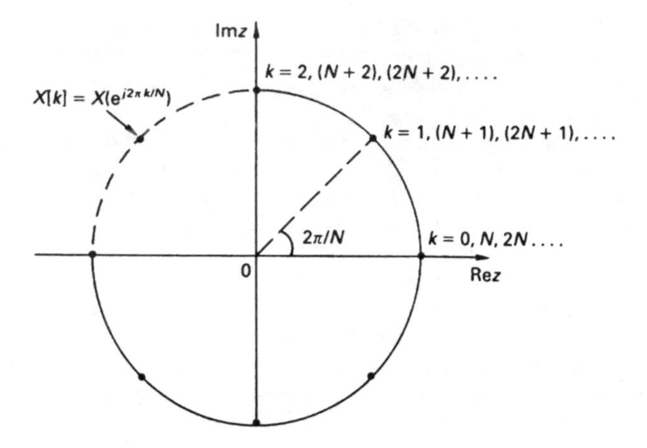

Fig. 7.15 The complex exponentials $e^{j2\pi k/N}$ has N distinct values which define N points equally spaced around the unit circle.

The N data values $X[k]$ are known collectively as the *discrete spectrum* of the finite-length sequence $x[n]$. We shall not be too concerned at this stage with the question of interpolating the graph of the continuous spectrum $X(e^{j\theta})$ from the N values of $X[k]$. The essential point is that the N spectral values contain all the information we need to work backwards and reconstruct the N values of the original sequence $x[n]$. This two-way transformation is indicated by

The existence of the DFT as a formal transformation satisfies the intuitive notion that a signal represented by N sample values in the time domain can be adequately represented by N spectral samples.

$$\underset{\text{time domain}}{\underset{\text{discrete}}{x[n]}} \xleftrightarrow{\text{DFT, } N} \underset{\text{Fourier domain}}{\underset{\text{discrete}}{X[k]}} \tag{7.32}$$

where 'DFT, N' denotes the N-point discrete Fourier transform.

The forward and inverse DFTs

To arrive at a definition of the forward transform, we summarize the series of operations outlined above. We thus begin with the DTFT

$$X(e^{j\theta}) = \sum_{n=0}^{N-1} x[n]\, e^{-jn\theta}$$

and set $\theta = 2\pi k/N$, giving

$$X[k] = \sum_{n=0}^{N-1} x[n]\, e^{-j(2\pi/N)nk}, \; k = 0, 1, 2, \ldots, N-1. \tag{7.33}$$

The notation of Equation 7.33 is so cumbersome that it is often easier to think in terms of sample values taken around the unit circle.

This is the form of the DFT *analysis equation* given in most textbooks. When using Equation 7.33, we must first decide on a value of k. Once we have done this, k is held constant while the sum of N terms is calculated. For instance, if we require $X[0]$ we set $k = 0$ in Equation 7.33, giving

$$X[0] = \sum_{n=0}^{N-1} x[n]\, e^{-j0} = \sum_{n=0}^{N-1} x[n].$$

Choosing $k = 1$, we obtain:

$$X[1] = \sum_{n=0}^{N-1} x[n]\, e^{-j(2\pi/N)n}$$

$$= x[0] + x[1]\, e^{-j(2\pi/N)} + x[2]\, e^{-j(4\pi/N)}$$

$$+ \ldots + x[n-1]\, e^{-j(2\pi(N-1)/N)}$$

and so forth.

When we use the DFT we are representing a sequence $x[n]$ as a superposition of N complex exponential component sequences. The N complex values $X[k]$ specify the magnitudes and phases of the complex exponential components of $x[n]$.

Worked Example 7.7

Demonstrate that the summation in Equation 7.33 defines a periodic sequence with period N.

Solution: We can show this by calculating values of $X[k]$ outside the interval, $0 \leq k \leq N-1$. For instance, setting $k = N$, gives:

$$X[N] = \sum_{n=0}^{N-1} x[n]\, e^{-j(2\pi/N)nN}.$$

The exponential term reduces to $e^{-j2\pi n} = 1$. Therefore

$$X[N] = \sum_{n=0}^{N-1} x[n]$$

which is identical to the value $X[0]$, obtained by setting $k = 0$ in Equation 7.33.
Similarly, if we replace k by $k + N$, we obtain:

$$X[k+N] = \sum_{n=0}^{N-1} x[n]\, e^{-j(2\pi/N)n(k+N)}$$

$$= \sum_{n=0}^{N-1} x[n]\, e^{-j(2\pi/N)nk}\, e^{-j2\pi n}.$$

Again, we have the factor $e^{-j2\pi n}$ so we obtain $X[k+N] = X[k]$. In other words, the summation generates only N distinct values which repeat indefinitely with period N.

To complete the picture we must identify the *inverse* operation which enables us to work from the discrete spectrum and reconstruct $x[n]$ over the range $0 \leq n \leq N-1$. Since the proof is rather tedious we shall omit the details and simply *define* the inverse DFT or IDFT by the *synthesis equation*:

$$x[n] = \frac{1}{N} \sum_{k=0}^{N-1} X[k]\, e^{j(2\pi/N)kn}, \; n = 0, 1, 2, \ldots, N-1. \tag{7.34}$$

It is important to understand the significance of the two defining equations. Worked Example 7.7 showed that if we use Equation 7.33 to compute additional values of $X[k]$ outside the range $0 \leq n \leq N-1$ they form a periodic spectral sequence. However, for the purposes of defining the DFT we used only the N values within the fundamental interval. This, we maintain, is sufficient to represent the N values of the original sequence, $x[n]$.

If we apply the inverse DFT to the discrete spectrum, Equation 7.34 yields

The synthesis equation shows how the sequence $x[n]$ can be synthesized or reconstituted for $0 \leq n \leq N-1$ by adding together N complex exponential components with magnitudes and phases specified by the complex values $X[k]$.

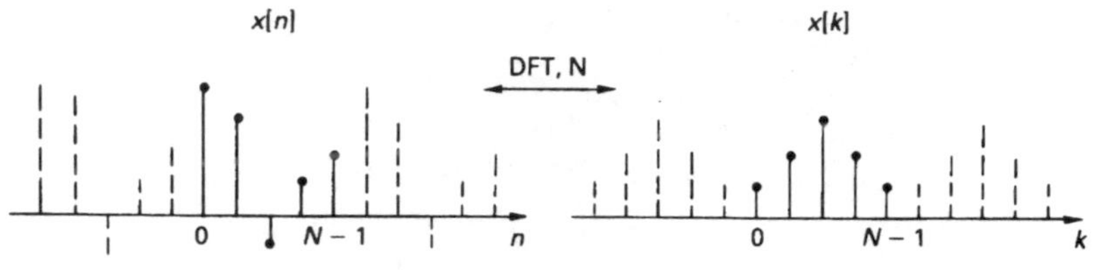

Fig. 7.16 The periodic nature of the forward and inverse DTFs.

$$x[n] = x_n, \; 0 \leq n \leq N - 1$$

as expected. But the summation in Equation 7.34 is, again, periodic. So, if we remove the restriction $0 \leq n \leq N - 1$, the inverse DFT generates an infinitely repeating pattern of sample values based on the original sequence $x[n]$, as illustrated in Figure 7.16. Notice that, as before, there are only N distinct values: the periodic extension of $x[n]$ provides no additional information beyond that contained in the fundamental interval $n = 0$ to $n = N - 1$.

By restricting the range of the inverse DFT to $0 \leq n \leq N - 1$ we are effectively representing a finite-length sequence as one period of an infinite periodic sequence.

Worked Example 7.7

a) Refer to Worked Example 7.1 and write down the DTFT, $X(e^{j\theta})$, of the Hanning sequence, $x[n] = \frac{1}{4}, \frac{1}{2}, \frac{1}{4}$; for $n = 0, 1, 2$.
b) By working from a graph of $X(e^{j\theta})$, construct a plot showing the magnitude and phase of the 3-point DFT, $X[k]$, corresponding to the original sequence, $x[n]$.
c) Use Equation 7.34 to invert the DFT and thereby recover $x[n]$ for $n = 0, 1, 2$.

Solution:

a) The Hanning sequence has the DTFT

$$X(e^{j\omega T}) = \tfrac{1}{2}(1 + \cos \omega T)\, e^{-j\omega T}.$$

Substituting $\theta = \omega T$, we obtain the spectrum in terms of $e^{j\theta}$:

$$X(e^{j\theta}) = \tfrac{1}{2}(1 + \cos \theta)\, e^{-j\theta}.$$

b) Figure 7.17a shows the graph of the magnitude and phase of $X(e^{j\theta})$, obtained by rescaling the horizontal axis in the original plot of Figure 7.3. The DFT of the three-term Hanning sequence is obtained by sampling the spectrum in the range $\theta = 0$ to $\theta = 2\pi$ at intervals of $2\pi/N$, with $N = 3$. We see from Figure 7.17b that the discrete spectrum is defined by the three values

$$X[0] = 1; \quad X[1] = \tfrac{1}{4}e^{-j2\pi/3}; \quad X[2] = \tfrac{1}{4}e^{j2\pi/3}$$

c) To recover the values of $x[n]$ from $X[k]$ we use the inverse DFT with $N = 3$:

$$x[n] = \frac{1}{3}\sum_{k=0}^{2} X[k]\, e^{j(2\pi/3)kn}, \quad n = 0, 1, 2.$$

To find $x[0]$, we set $n = 0$, giving

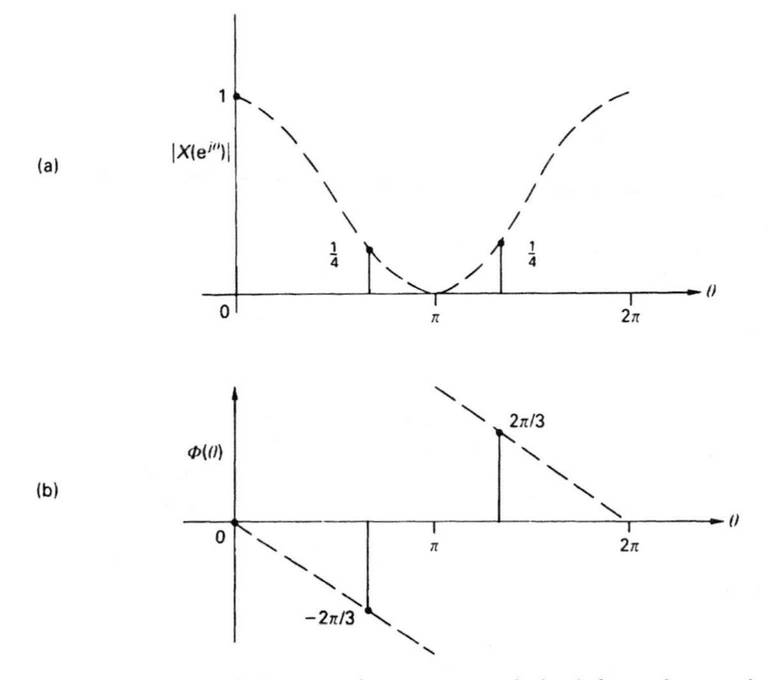

Fig. 7.17 The DFT of the Hanning sequence derived from the graph of its DTFT.

$$\begin{aligned} x[0] &= (X[0] + X[1] + X[2])/3 \\ &= (1 + \tfrac{1}{4}e^{-j2\pi/3} + \tfrac{1}{4}e^{j2\pi/3})/3 \\ &= (1 + \tfrac{1}{2}\cos 2\pi/3)/3 \\ &= \tfrac{1}{4}. \end{aligned}$$

Notice that we have remembered to include the factor of $1/N$ which appears in the inverse DFT!

Next we evaluate the summation with $n = 1$:

$$\begin{aligned} x[1] &= \frac{1}{3}\sum_{k=0}^{2} X[k]\,e^{j(2\pi/3)k} \\ &= (X[0] + X[1]e^{j(2\pi/3)} + X[2]e^{j(4\pi/3)})/3 \\ &= (1 + \tfrac{1}{4} + \tfrac{1}{4}e^{j2\pi})/3 \\ &= \tfrac{1}{2}. \end{aligned}$$

Finally, with $n = 2$:

$$\begin{aligned} x[2] &= \frac{1}{3}\sum_{k=0}^{2} X[k]\,e^{j(4\pi/3)k} \\ &= (X[0] + X[1]e^{j(4\pi/3)} + X[2]e^{j(8\pi/3)})/3 \\ &= (1 + \tfrac{1}{4}e^{j2\pi/3} + \tfrac{1}{4}e^{j10\pi/3})/3 \end{aligned}$$

Now, $e^{j10\pi/3} = e^{j4\pi}e^{-j2\pi/3} = e^{-j2\pi/3}$. The final part of the calculation is therefore similar to that shown for the case $k = 0$, giving $x[2] = \frac{1}{4}$.

We saved time and effort in the previous example by obtaining the values of the discrete spectrum $X[k]$ directly from the continuous function $X(e^{j\theta})$. Otherwise, if we had used the forward transform, as defined by Equation 7.33, the calculation of $X[k]$ would have been no less tedious than the inverse calculation, whereby we recovered the sequence $x[n]$ from the spectral samples. It goes without saying that the amount of work involved in such calculations increases quite dramatically as the number of sample values increases. However, it also becomes apparent that there is a certain amount of redundancy in the calculation due to the fact that the complex exponential $e^{j(2\pi/N)nk}$ is a periodic function which takes only N distinct values. As a result, we find that the same values $x[n]\,e^{j(2\pi/N)nk}$ are used many times, the more so as the value of N becomes larger.

It is fairly easy to show that when the DFT of an N-point sequence is calculated directly from Equation 7.33 the number of multiplications involved increases as N^2.

When the DFT is computed using an FFT algorithm the number of multiplications is reduced to the order of $\log_2 N$.

An FFT algorithm that operates on a sequence of length $N = 2^A$, where A is integer, is sometimes known as a *radix-2* algorithm.

The class of FFT algorithms mentioned earlier was developed with the aim of eliminating much of this redundancy. It is worth pointing out that the derivation of the FFT illustrates a 'divide-and-conquer' approach to problem-solving in which the computation of a large N-point DFT is reduced to the computation of smaller DFTs which can be stored and later combined to produce the larger DFT. Most, though by no means all, of the FFT algorithms in common use require that the number of data points N is an integer power of 2.

Summary

We have covered a considerable amount of ground in this chapter. The starting point was to show that the steady-state analysis of discrete-time systems can be carried out using frequency-domain models analogous to those that we first met in Chapter 3, in the context of continuous-time signals and systems. In the discrete-time case the application of the discrete-time Fourier transform or DTFT transforms the unit-sample response $h[n]$ of a system to the periodic frequency-response function $H(e^{j\omega T})$. When applied to an aperiodic sequence $x[n]$ the DTFT gives the spectrum of the sequence $X(e^{j\omega T})$. The spectrum is a periodic function of frequency which specifies the magnitude and phase of the sinusoidal component sequences of $x[n]$ distributed over the frequency axis.

The introductory sections emphasized the essential similarities between the use of the Laplace and z-transforms in their respective domains and further highlighted the relationship between pole-zero descriptions of signals and systems and their Fourier descriptions in the frequency domain. We saw in particular that the frequency-response of a discrete-time system and the spectrum of a sequence can be found by evaluating the transfer function $H(z)$ or the z-transform $X(z)$ on and around the unit circle in the complex z-plane.

We noted the close correspondence between the properties of the DTFT and the continuous-time Fourier transform. Because the spectrum of an aperiodic sequence is periodic in frequency, the spectrum can be inverted using the techniques of Fourier analysis, with a view to recovering the values of the original sequence.

We then turned our attention to the problem of relating the spectrum of a continuous-time signal $x(t)$ to the spectrum of its samples. We showed that if $x(t)$ was bandlimited and properly sampled following the sampling rule, its spectrum could be deduced from a knowledge of its sample-data sequence.

If a signal is truncated to a finite length its spectrum can be calculated from a finite-length sequence, but the frequency resolution that can be achieved is then limited to the reciprocal of the record length. This means in turn that the spectrum can be specified quite adequately by calculating a finite number of spectral samples equally spaced over the interval $0 \leqslant \omega < \omega_s$. The number of spectral samples required is equal to the number of values N in the corresponding sample-data sequence. The minimum number is given when the sampling is carried out at the lowest possible frequency consistent with the sampling frequency.

The calculation of spectral samples was considered in further detail and we showed that the values of a sequence $x[n]$ are uniquely related to the samples of its spectrum by the discrete Fourier transform or DFT.

We can trace the developments leading up to the DFT by working through Figure 7.18. The diagram shows a number of different types of signal together with the spectral description that is appropriate in each case. Beginning at the top of the diagram, we have an aperiodic signal $x(t)$ related to its continuous spectrum $X(j\omega)$ by the continuous-time Fourier transform. We recall that the spectrum is an *aperiodic* continuous function of frequency.

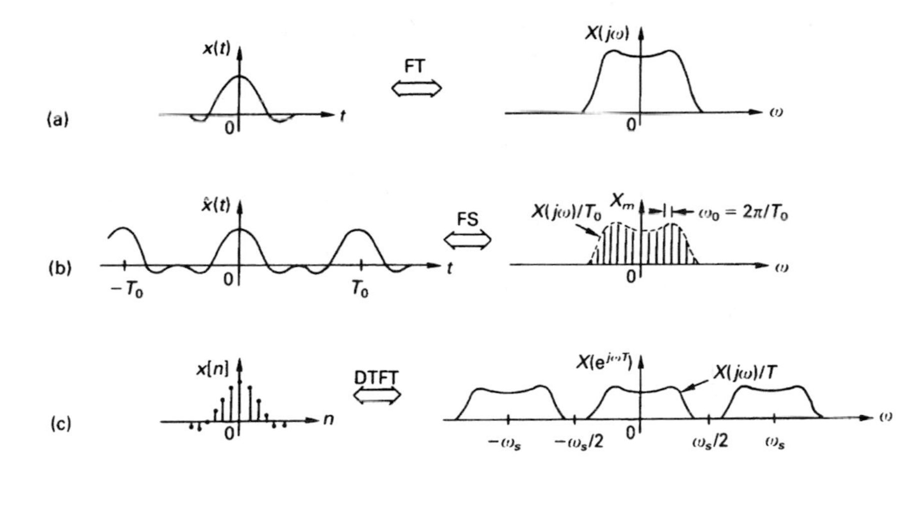

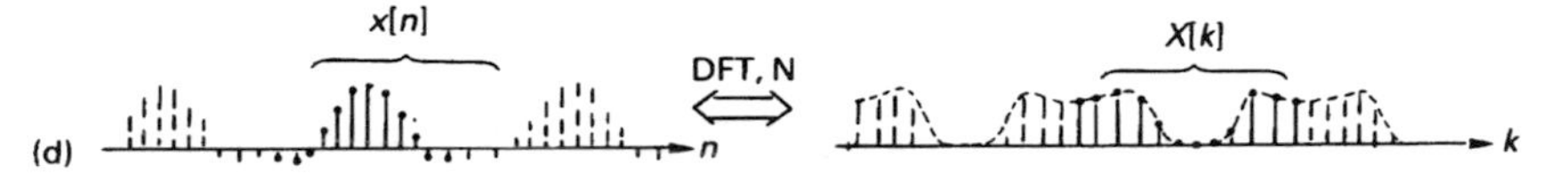

Fig. 7.18 Different types of signal and their transformations. (a) A continuous-time aperiodic signal and its Fourier transform; (b) a continuous-time periodic signal and its discrete Fourier spectrum; (c) an aperiodic discrete-time sequence with its DTFT; (d) a finite length discrete-time sequence and its discrete spectrum given by the DFT.

Next, we see a continuous-time *periodic* signal $\hat{x}(t)$ formed by adding $x(t)$ to its displacements of the form $x(t - kT_0)$ along the time axis. The signal is described in terms of a harmonic Fourier series and has an aperiodic *discrete* spectrum with the Fourier coefficients defined by the relationship $X_m = X(jm\omega)/T_0$.

The discrete-time signal in Figure 7.18c has been formed by sampling $x(t)$ at the instants $t = nT$, where n is integer. Although the signal is defined by a discrete-time sequence, it is associated with a *continuous* spectrum $X(e^{j\omega T})$ given by the continuous-time Fourier transform, the DTFT. We have seen that the spectrum is periodic in frequency, repeating over an interval equal to the sampling frequency $\omega_s = 2\pi/T$. Because $x[n]$ has been derived from $x(t)$, we can find the spectrum of the samples by adding together shifted spectra of the form $X(j\omega - k\omega_s)/T$. In the example shown, the sampling frequency has been chosen so that $\omega_s > 2f_B$ where f_B is the signal bandwidth. This means that the sampling frequency has been obeyed with the result that the spectrum of $x(t)$ is unambiguously related to the spectrum of $x[n]$.

Finally, in Figure 7.18d, the N values of $x[n]$ have been transformed using an N-point DFT to obtain the N values of the discrete spectrum $X[k]$, equally spaced over the interval $0 \leqslant \omega < \omega_s$. The N values of the original sequence and the N calculated spectral values (shown with solid lines in the figure) are uniquely related by the DFT. The dotted lines in the discrete spectrum indicate the periodic extension of $X[k]$ that would be found by calculating the forward transform for $k < 0$ or $k \geqslant N$. Similarly, if the inverse DFT is computed for $n < 0$ or $n \geqslant N$, we simply regenerate the values of $x[n]$ and so obtain an N-periodic sequence based on the original sequence. We thus conclude that the DFT treats an aperiodic sequence as if it were periodic for the purposes of calculation.

We commented briefly on the use of FFT algorithms for efficient computation of the DFT. An FFT-based software package transforms a general-purpose digital computer into a powerful signal processing system. It should be evident, nevertheless, that the value of such a system is greatly enhanced if the user is aware of the analytical background and able to make informed judgements with regard to signal handling, signal truncation and the effects of working with a finite frequency resolution.

Problems

7.1 Find the DTFT of the sequence $x[n] = 1, 0, 1$ for $n = 0, 1, 2$.

7.2 The sum of the first N terms of a geometric series is given by

$$a + ar + ar^2 + \ldots + ar^{N-1} = \frac{a(1 - r^N)}{1 - r}$$

for $r \neq 1$.

Use this result to obtain a closed-form expression for the DTFT of the finite-length sequence $x[n]$ defined by

$$x[n] = \alpha^n \quad \text{for } 0 < n \leqslant N - 1$$

Formulae for calculating sums of series are given in Szymanski, J.E., *Basic Mathematics for Electronic Engineers*, Tutorial Guides in Electronic Engineering, No. 16, Chapman and Hall.

7.3 Using the formula for the partial sum of a geometric series given in problem 7.2, find a general closed-form expression for the DTFT of the N-point sequence of unit sample values:

Some work with trigonometric identities may be required in the second part of this question; the answer is not obvious at first sight!

$$x[n] = u[n] - u[n - N].$$

Show that if $N = 3$, the general result reduces to the DTFT of the 3-point sequence considered in Worked Example 7.4.

7.4 A discrete-time linear phase *all-pass* system has a frequency-response function $H(e^{j\omega T})$ with magnitude

$$|H(e^{j\omega T})| = 1 \quad \text{for } |\omega| \leq \omega_s/2$$

and phase

$$\Phi(\omega) = -2\omega T.$$

Determine the response of this system to the discrete-time sequence $x[n]$ shown opposite.

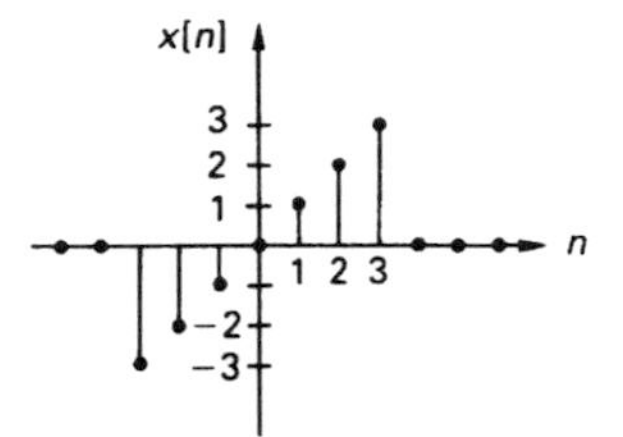

7.5 The discrete-time sequence from problem 7.4 has the DTFT

$$X(e^{j\theta}) = |X(e^{j\theta})|e^{j\Phi(\theta)}.$$

Obtain numerical values for the following quantities, *without* first calculating the DTFT of the sequence:

a) $X(e^{j\theta})$ for $\theta = 0$;
b) $\Phi(\theta)$
c) $X(e^{j\pi})$
d) $\int_{-\pi}^{\pi} X(e^{j\theta})\, d\theta/2\pi$

7.6 A non-causal sequence has values $x[n] = (2\sin n\pi/2)/n\pi$ for $|n| < 5$ and is zero elsewhere. Sketch its spectrum for $|\omega| < \omega_s/2$ using known results for the Fourier time series.

7.7 Find the sequence of values associated with a spectrum of the form $X(e^{j\theta}) = (1 + \cos\theta + \frac{1}{2}\cos 2\theta + \frac{1}{3}\cos 3\theta)\, e^{-j3\theta}$.

7.8 Refer to Worked Example 7.2 and, hence, find the unit-sample response of a discrete-time system characterized by the frequency-response function

$$H(e^{j\theta}) = \frac{e^{-j\theta} - \alpha}{1 - \alpha e^{-j\theta}}.$$

7.9 A signal record of duration 25 s is sampled at 4096 equally-spaced points and the discrete spectrum of the sample sequence is computed using the DFT. Calculate a) the spacing (in Hz) of the frequency components, and b) the highest frequency present in the discrete spectrum.

7.10 A signal is effectively bandlimited to 20 kHz. Its spectrum is to be found using the DFT with a resolution of 100 Hz. The FFT algorithm used in the computation requires that the number of data values N is an integer power of 2. What is the minimum number of samples required?

Appendix

A short table of Laplace transform pairs

Time function $f(t)$	Laplace transform $F(s)$
$\delta(t)$	1
$u(t)$	$1/s$
t	$1/s^2$
t^n	$n!/s^{n+1}$
$e^{-\alpha t}$	$1/(s+\alpha)$
$t e^{-\alpha t}$	$1/(s+\alpha)^2$
$t^n e^{-\alpha t}/n!$	$1/(s+\alpha)^{n+1}$
$\sin \omega t$	$\omega/(s^2+\omega^2)$
$\cos \omega t$	$s/(s^2+\omega^2)$
$e^{-\alpha t} \sin \omega t$	$\omega/[(s+\alpha)^2+\omega^2]$
$e^{-\alpha t} \cos \omega t$	$(s+\alpha)/[(s+\alpha)^2+\omega^2]$

Some Laplace transform properties

Time domain		s-domain
$f(t)$	$\leftrightarrow$	$F(s)$
$f(\alpha t)$	$\leftrightarrow$	$(1/\alpha)F(s/\alpha)$ for $\alpha > 0$
$f(t-T)u(t-T)$	$\leftrightarrow$	$e^{-sT}F(s)$
$f(t)e^{-\alpha t}$	$\leftrightarrow$	$F(s+\alpha)$
$df(t)/dt$	$\leftrightarrow$	$sF(s) - f(0)$
$d^2f(t)/dt^2$	$\leftrightarrow$	$s^2F(s) - sf(0) - f'(0)$
$d^nf(t)/dt^n$	$\leftrightarrow$	$s^nF(s) - s^{n-1}f(0) - s^{n-2}f'(0) - \ldots - f^{(n-1)}(0)$
$\int_0^t f(\tau)\,d\tau$	$\leftrightarrow$	$F(s)/s$
$\int_0^t x(\tau)h(t-\tau)\,d\tau$	$\leftrightarrow$	$X(s)\cdot H(s)$

Some z-transform pairs

Sequence $f[n]$ ($f[n] = 0$ for $n < 0$)	z-transform $F(z)$
$\delta[n]$	1
$u[n]$	$\frac{z}{z-1}$
n	$\frac{z}{(z-1)^2}$
a^n	$\frac{z}{z-a}$
e^{-anT}	$\frac{z}{z-e^{-aT}}$
na^n	$\frac{z}{(z-a)^2}$
$\sin n\omega T$	$\frac{z\sin\omega T}{z^2 - 2z\cos\omega T + 1}$
$\cos n\omega T$	$\frac{z^2 - z\cos\omega T}{z^2 - 2z\cos\omega T + 1}$

Answers to Numerical Problems

1.1 $a[n] = 0, 1, 2, 3, \ldots$ and $b[n] = 1, 2, 4, 8, \ldots$ for $n = 0, 1, 2, 3 \ldots$ Thus, $a[n] + b[n] = 1, 3, 6, 11, \ldots$

1.8 a) $N = 4$; b) $N = 8$

1.10 The number of samples N taken over a signal record of length T_0 will be $N = f_s T_0$. Since the maximum frequency content of the signal is 250 Hz, f_s must be chosen so that $f_s > 500$ Hz, giving $N > 500\,\text{Hz} \times 100\,\text{ms}$, i.e. $N > 50$. Thus if $N = 60$ as in option c) there will be sufficient samples to represent the signal.

2.1 (a) $y[n] = 1/2, 1, 1, 1, 1, 1, \ldots$
(b) $y[n] = 1, 3/2, 7/4, 15/8, 31/16, 63/32, \ldots$

2.2 (a) $y[n] = 1/2, 1, 1, 1, 1/2, 0, 0, 0, \ldots$
(b) $y[n] = 1, 3/2, 7/4, 15/8, 15/16, 15/32, 15/64, \ldots$
The response of the non-recursive processor (a) is a finite-length sequence while the response of the recursive processor (b) is an infinite sequence.

2.3 $y[2] = -2.82$, $y[5] = -7.61 \times 10^{-2}$.

2.4 $y(t) = -0.717$ at $t = 1.5$ s.

2.5 2.

2.6 4.95×10^{-2} V.

2.7 The output can be expressed as the sum of a steady sinusoid and a decaying exponential

$$y(t) = \frac{\cos(\omega_0 t - \theta)}{\sqrt{(\alpha^2 + \omega_0^2)}} - \frac{\alpha e^{-\alpha t}}{(\alpha^2 + \omega_0^2)}$$

where $\theta = \tan^{-1}(\omega_0/\alpha)$.
For $t \gg 1/\alpha$ the decaying exponential, or transient, part of the response will be negligibly small. Hence $y(t)$ will be equal to the steady-state sinusoidal response of the network.

2.8 The unit-sample response for both processors is the finite sequence

$$h[n] = 1, 0.5, 0, 0, 0, \ldots$$

Note that a recursive realisation, implied by the linear difference equation (ii), does not necessarily lead to an infinite impulse response.

2.9 The output sequence is

$$y[n] = \ldots, 0.4, 0.5, -0.7, -0.25, 0.55, 0.65, -0.9, 0.65, 0.25, \ldots$$

This processor *accentuates* changes in the input sequence, unlike the moving averager which tends to smooth out variations. Because the processor responds to the difference between $x[n]$ and $x[n-1]$, small changes between samples lead to low outputs while large changes produce large outputs. The sign of the output reflects the direction of the change.

2.10 The output sequence may be written in the form

$y[n] = A\cos(n-1)\,\omega_0 T$, where $A = |1 + \cos\omega_0 T|$.

3.1 $v(t) = (1.5\,e^{j\pi/4})\,e^{-j4t} + 0.5\,e^{-j2t} + 3 + 0.5\,e^{j2t} + (1.5\,e^{-j\pi/4})\,e^{j4t}$.
3.2 (i) $y(t) = 6.71\sin(0.5t - 0.58)$.
(ii) $0.25\,\text{rad s}^{-1}$, $-45°$ or $-\pi/4$ rad.
3.3 $h(t) = 0.75\,e^{-t/4}$.
3.4 $H(j\omega) = (j\omega + 5)/(j\omega + 2)(j\omega + 3)$.
3.6 $\delta(t-\tau) \leftrightarrow e^{-j\omega\tau}$; $|X(j\omega)| = 2|1 + \cos\omega\tau|$. The spectrum is periodic in frequency, repeating at intervals of $2\pi/\tau$.
3.7 $H(j\omega) = (6 \times 10^6)/((j\omega)^2 + 1600(j\omega) + 10^6)$; $1000\,\text{rad s}^{-1}$.

3.8 $$V(j\omega) = \frac{VT}{2}\left[\frac{\sin(\omega - \omega_0)T/2}{(\omega - \omega_0)T/2} + \frac{\sin(\omega + \omega_0)T/2}{(\omega + \omega_0)T/2}\right]$$

where $V = 5$, $T = 20$ ms and $\omega_0 = 2000\pi\,\text{rad s}^{-1}$.
3.9 $h(t) = (\omega_0/\pi)\sin(\omega_0 t)/\omega_0 t$.
The impulse response is not causal; it begins before the impulse is applied at $t = 0$. This is impossible, hence the ideal filter is physically unrealizable.

3.10 $$X(j\omega) = \frac{2\pi^2 \sin\omega\tau/2}{\omega(\pi^2 - (\omega\tau/2)^2)}.$$

$X(0) = \tau$, and the first zero-crossings occur at $\omega = \pm 4\pi/\tau$.
4.1 (a) $(3s + 4)/(s + 1)(s + 2)$.
(b) $(e^{-3s})/s^2$.
(c) $(s + 3)/\sqrt{2}(s^2 + 9)$.
(d) $(s + 3)/(s^2 + 6s + 25)$.
4.2 (a) $(\frac{1}{2})u(t) + (\frac{1}{2})e^{-2t}$.
(b) $e^{-3t}u(t) - e^{-3(t-1)}u(t-1)$.
(c) $(e^{-t}\sin 3t)/3$.
(d) $(t^2/2)\,e^{-4t}$.
4.3 (a) $I(s) = sCV(s) - Cv(0)$.
(b) $V(s) = sLI(s) - Li(0)$.
4.4 (a) $v_0(t) = (1 - e^{-100t})$ for $0 \leq t \leq 0.1$ s
$= (1 - e^{-100t}) - (1 - e^{-100(t-0.1)})$ for $t \geq 0.1$ s.
(b) $v_0(t) = 50(e^{-20t} - e^{-100t})/8$.
4.5 $f(t) = (1 + t)\,e^{-2t}$.
4.6 zero: $s = -1/4$, poles: $s = -2 \pm j3$;
$h(t) = 4.63\,e^{-2t}\sin(3t + 2.1)$.
4.7 (a) poles: $s = -1/\sqrt{2} \pm j1/\sqrt{2}$.
(b) zeros: $s = \pm j$, double pole at $s = -1$.
(c) zeros: $s = \pm j\sqrt{2}$, poles: $s = -1$, $s = -0.1 \pm j0.995$.
(d) zeros: $s = +1/2 \pm j\sqrt{3}/2$, poles: $s = -1/2 \pm j\sqrt{3}/2$.
4.8 Closed-loop pole positions

for $K = 0$: $s = -(3 \pm \sqrt{5})/2$
for $K = 1$: double pole at $s = -1$
for $K = 2$: $s = -1/2 \pm j\sqrt{3}/2$
for $K = 3$: $s = \pm j$
for $K = 4$: $s = 1/2 \pm j\sqrt{3}/2$

System is unstable for $K > 3$.
4.9 $V(s)/I(s) = (s + 5)/(s^2 + 3s + 4)$; steady-state output voltage $v(t)$ will be

a sinewave of frequency 1 rad s^{-1}, amplitude 1.2 mV and phase (relative to the input) of $-33.7°$.

4.10 $y(t) = u(t) - 3e^{-2t} + 2e^{-3t}$. $y(t)$ reaches 0.95 in approximately 2 seconds and approaches a final value of 1 as $t \to \infty$.

5.1 $H(z) = 2 - 5z^{-1} - z^{-2} + 6z^{-3}$
$h[n] = 2, -5, -1, 6, 0, 0, 0, \ldots$

5.2 The output is the sequence of values $y[n] = 1/N, 2/N, 3/N, \ldots$ which reaches the steady-state value of unity in N terms, or $N - 1$ sample periods.

5.3 The output is a sequence of values of alternating sign:

$$y[n] = 1, -1/4, 1/16, -1/64, 1/256, \ldots$$

5.4 (a) $X(z) = 2z/(z - 1)^2$.
(b) $X(z) = 5z/(z - e^{-0.2}) = 5z/(z - 0.819)$.
(c) $X(z) = 3z(1 - e^{-0.5})/(z - 1)(z - e^{-0.5})$
$= 1.18z/(z - 1)(z - 0.607)$.

5.5 (a) 1, 2, 3, 0, 0, . . .
(b) 3, 1, 1/3, 1/9, 1/27, . . .
(c) 0, 1, 1.9, 2.71, 3.44, . . .
(d) 0, 0, 0, 0.8, 1.28, 1.56, 1.64, 1.64, 1.01, . . .

5.6 The unit-step response can be expressed as the sum of three infinite sequences:

$$y[n] = 1.563u[n] + 0.188(-0.6)^n - 0.75(0.6)^n$$

where $u[n]$ represents the unit-step sequence.

5.7 (a) The transfer function is $H(z) = z(z + 1)/(4z^2 + 1)$. The zeros lie at $z = 0$ and $z = -1$, and the poles lie at $z = \pm j0.5$. Since all the poles lie within the unit circle the system will be *stable*.
(b) The transfer function can be written as $H(z) = z(z^2 - 2)/(z - 1)(z^2 + 2)$. The zeros lie at $z = 0$, and $z = \pm\sqrt{2}$, and the poles lie at $z = 1$ and $z = \pm j\sqrt{2}$. Since two poles lie outside of the unit circle the system will be *unstable*.

5.8 Owing to the zeros at $z = \pm 1$, the response is zero at the frequencies $\omega = 0$ and $\omega = \omega_s/2$. The magnitude response peaks at $z = e^{\pm j\omega_0 T}$, in the neighbourhood of the poles, where $\omega_0 T = \pi - \tan^{-1}(0.814/0.47) = 2.094$ rad. Since $\omega_0 = 2\pi f_0$ and $T = 1/f_s$, we have $\omega_0 T = 2\pi f_0/f_s$. For a sampling frequency $f_s = 12$ kHz, the centre frequency $f_0 = 2.094 \times f_s/2\pi = 4$ kHz.

5.10 $|H(e^{j\omega T})| = |(1 + 2\cos\omega T)/3|$. The magnitude response is 1, 0.127 and 0.333 at the frequencies 0, 300 Hz and 500 Hz respectively, and zero at 333 Hz.

5.11 At 10 Hz the magnitude of the frequency response function is 0.342 and the phase shift is -1.55 rad (about $-89°$). Hence the output component at 10 Hz will have an amplitude of $0.5 \times 0.342 = 0.171$ V, and will lag the input by 1.55 rad, or about 87°. The system will have an identical response to frequency components at 10 Hz, 35 Hz, 60 Hz and so on.

6.1 $x(t) = 1 + 2\cos 2\omega_0 t + 3\sin 3\omega_0 t$, where $\omega_0 = 5$; $T_0 = 2\pi/5$. $X_0 = 1$, $X_2 = X_{-2} = 1$, $X_3 = -j3/2$, $X_{-3} = j3/2$.

6.2 $P_{av.} = 1 + 2^2/2 + 3^2/2 = 7.5\,\text{W}$.

6.3 $X_m = j(-1)^m/m\pi$.

6.4 $|X_m| = 10^{-4}\omega_c^2/(m^2\omega_0^2 + \omega_c^2)\,T_0$; $\theta_m = -2\tan^{-1} m\omega_0/\omega_c$. The system has an impulse response, $h(t) = \omega_c^2 t e^{-\omega_c t}$, hence pulse separation should be chosen so that $T_0 > 5/\omega_c$, i.e. $T_0 > 5.10^{-3}\,\text{s}$.

7.1 $X(e^{j\omega T}) = (2\cos\omega T)\,e^{-j\omega T}$.

7.2 In terms of the geometric series, $a = 1$ and $r = \alpha e^{-j\omega T}$. The DTFT is therefore

$$X(e^{j\omega T}) = \frac{(1 - \alpha^N e^{-jN\omega T})}{1 - \alpha e^{-j\omega T}}.$$

7.3 $x[n]$ is a sequence of N samples, each of unity value. The DTFT can therefore be found by setting $\alpha = 1$ in the DTFT of the previous problem, giving

$$\begin{aligned} X(e^{j\omega T}) &= \frac{(1 - e^{-jN\omega T})}{1 - e^{-j\omega T}} \\ &= \frac{(e^{jN\omega T/2} - e^{-jN\omega T/2})\,e^{-jN\omega T/2}}{(e^{j\omega T/2} - e^{-j\omega T/2})\,e^{-j\omega T/2}} \\ &= \frac{\sin(N\omega T/2)}{\sin(\omega T/2)} e^{-j(N-1)\omega T/2}. \end{aligned}$$

7.4 The frequency response of the all-pass discrete-time system can be written as $H(e^{j\omega T}) = e^{-2j\omega T}$ for $|\omega| < \omega_s/2$. Its only effect therefore will be to *delay* an input sequence by two sampling intervals. The response to the sequence $x[n]$ is therefore the delayed sequence $y[n] = x[n - 2]$.

7.5 The sequence is an *odd* function of n. Its DTFT is therefore *imaginary* and given by a Fourier *sine* series as indicated in exercise 7.3(a). It therefore follows that:

(a) If $\theta = \omega T = 0$, then $X(e^{j\theta}) = 0$.

(b) $e^{j\Phi(\omega)} = -j$, so $\Phi(\omega) = -\pi/2$.

(c) If $\theta = \omega T = \pi$ then $X(e^{j\theta}) = 0$, since $\sin n\pi = 0$, for integer n.

(d) The integral is the defining integral for the inverse DTFT evaluated for $n = 0$. Therefore:

$$\int_{-\pi}^{\pi} X(e^{j\theta})\,d\theta/2\pi = x[0] = 0.$$

7.7 The DTFT corresponds to the causal sequence $x[n] = \frac{1}{6}, \frac{1}{4}, \frac{1}{2}, 1, \frac{1}{2}, \frac{1}{4}, \frac{1}{6}$.

7.8 The frequency response function can be expressed as

$$H(e^{j\theta}) = \frac{e^{j\theta}}{e^{j\theta} - \alpha} e^{-j\theta} - \frac{\alpha e^{j\theta}}{e^{j\theta} - \alpha}$$

which is the difference between the DTFTs of two sequences $h[n - 1]$ and $\alpha h[n]$, where $h[n] = \alpha^n$ for $n \geqslant 0$.

7.9 The spectral spacing is equal to the frequency resolution given by the reciprocal of the record length, thus $\delta f = 1/25 = 0.04\,\text{Hz}$.

The highest frequency represented by an N-point DFT is $(N - 1)\delta f$. Since $N = 4096$, the highest frequency is $4095 \times 0.04 = 163.8\,\text{Hz}$.

7.10 We have $T_0 = N/f_s$, where N is to be an integer power of 2. To achieve a frequency resolution of 100 Hz we require $T_0 \geqslant 1/100$ s. Thus,

$$T_0 = N/f_s = 2^A/f_s \geqslant 0.01\,\text{s}.$$

If $f_s = 2f_B$, then $f_s = 40\,\text{kHz}$, giving $2^A \geqslant 0.01 \times 40 \times 10^3$. The smallest value of A satisfying this condition is 9, when $N = 2^A = 512$.

Index

Printed in the United States
120963LV00005B/171-180/A

9 780412 401107